Praise for

PROOF

"A romp through the world of alcohol." —***New York Post***

"Wine lovers, beer hounds, whiskey connoisseurs, and even teetotalers are all likely to find something to interest them in this look at the science of liquor." —***Scientific American***

"A great read for barflies and know-it-alls — or the grad student who is likely both." —***New York Times T Magazine***

"In this brisk dive into the history and geekery of our favorite social lubricant, *Wired* editor Adam Rogers gets under the cap and between the molecules to show what makes our favorite firewaters so irresistible and hard to replicate — and how a good stiff drink often doubles as a miracle of human ingenuity." —***Mother Jones***

"A comprehensive, funny look at booze . . . Like the best of its subject matter, *Proof*'s blend of disparate ingredients goes down smooth, and makes you feel like an expert on the topic." —***Discover***

"This science-steeped tale of humanity's 10,000-year love affair with alcohol is an engaging trawl through fermentation, distillation, perception of taste and smell, and the biological responses of humans to booze . . . *Proof* is an entertaining, well-researched piece of popular-science writing." —***Nature***

"A whiskey nerd's delight . . . Full of tasty asides and surprising science, this is entertaining even if you're the type who always drinks what the other guy is having." — ***Chicago Tribune***

"Rogers tackles drinking from his particular prism — the mechanics of how it's made and how it affects us, everything from fermentation to physical effects. It's testament to Rogers's skills (he's the articles editor for *Wired*) that he weaves deep research into an engaging, smart read." — ***San Francisco Chronicle***

"Written in the same approachable yet science-savvy tone of other geeky tomes (think Amy Stewart's *The Drunken Botanist* and Brian Greene's *The Fabric of the Cosmos*), Rogers' book sheds light on everything from barrels to bacteria strains." — ***Imbibe***

"A spirited book . . . An easy-reading, enjoyable book that does a great job of surveying both the history and research of what's in your glass . . . Delicious, tasty science." — ***Wine Spectator***

"This paean to booze is a thought-provoking scientific accompaniment to your next cup of good cheer." — ***The Scientist***

"Sharp, entertaining, informative distillation." — ***Newsday***

"Follow a single, microscopic yeast cell down a rabbit hole, and Alice, aka Adam, will take you on a fascinating romp through the Wonderland of ethyl alcohol, from Nature's own fermentation to today's best Scotch whiskies — and worst hangovers. This book is a delightful marriage of scholarship and fun." — **Robert L. Wolke, author of *What Einstein Kept Under His Hat* and *What Einstein Told His Cook***

"*Proof,* this irresistible book from Adam Rogers, shines like the deep gold of good whiskey. By which I mean it's smart in its science, fascinating in its complicated and very human history, and entertaining on all counts. And that it will make that drink in your hand a lot more interesting than you expected." **—Deborah Blum, author of *The Poisoner's Handbook: Murder and the Birth of Forensic Medicine in Jazz Age New York***

"Absolutely compelling. *Proof* sits next to Wayne Curtis's *And a Bottle of Rum* and Tom Standage's *A History of the World in Six Glasses* as a must-read." **—Jeffrey Morgenthaler, bar manager at Clyde Common and author of *The Bar Book***

"*Proof* is science writing at its best—witty, elegant, and abrim with engrossing reporting that takes you to the frontiers of booze, and the people who craft it."

—Clive Thompson, author of *Smarter Than You Think*

"Rogers distills history, archaeology, biology, sociology, and physics into something clear and powerful, like spirits themselves."

—Jim Meehan, author of *The PDT Cocktail Book*

"A page-turner for science-thirsty geeks and drink connoisseurs alike, *Proof* is overflowing with fun facts and quirky details. I'm drunk—on knowledge!" **—Jeff Potter, author of *Cooking for Geeks***

"Adam Rogers writes masterfully and gracefully about all the sciences that swirl around spirits, from the biology of a hangover to the paleontology of microbes that transform plant juices into alcohol. A book to be savored and revisited."

—Carl Zimmer, author of *Parasite Rex* and *A Planet of Viruses*

"Reading *Proof* feels just like you're having a drink with a knowledgeable and enthusiastic friend. Rogers's deep affinity for getting to the bottom of his subject shines through on every page."

—Adam Savage, TV host and producer of *MythBusters*

"As a distiller I find most books on booze to be diluted. The science and history here are sure to satisfy the geekiest of drinkers, while the chapters, carried by stories, told through the lens of a rocks glass do not lose the casual. To get this kind of in-depth overview of how spirits are produced, consumed, and studied you'd have to read twenty books."

—Vince Oleson, Head Distiller/Barrel Thief, Widow Jane Distillery

"An entertaining read . . . Rogers elegantly charges through what took me more than five years of research to learn . . . *Proof* will inspire and educate the oncoming hordes who intend to make their own booze and tear down the once solid regulatory walls of the reigning royal houses of liquor." **—Dan Garrison, Garrison Brothers Distillery**

"The perfect book to enjoy with a glass of red." —***Guardian***

"An entertaining and enlightening nonfiction search for the science, art, and magic that lead to 'perfect bar moments.'" **—BBC *Culture***

"[Rogers] has a light, witty touch that keeps even the most scientific passages flowing like good drink." —***Federalist***

PROOF

PROOF

The SCIENCE *of* BOOZE

Adam Rogers

MARINER BOOKS
HOUGHTON MIFFLIN HARCOURT
BOSTON • NEW YORK

First Mariner Books edition 2015

www.hmhco.com

Library of Congress Cataloging-in-Publication Data
Rogers, Adam, date.
Proof : the science of booze / Adam Rogers.
pages cm
Includes bibliographical references and index.
ISBN 978-0-547-89796-7 (hardcover) ISBN 978-0-544-53854-2 (pbk.)
1. Liquors. 2. Alcoholic beverages. 3. Distillation. I. Title.
TP505.R64 2014
663'.1—dc23
2013045770

Book design by Brian Moore
Printed in the United States of America
DOC 10 9 8 7 6 5 4 3 2 1

FOR MELISSA

CONTENTS

INTRODUCTION

Deep in New York's Chinatown is a storefront made nearly invisible by crafty urban camouflage. The sign says that the place is an interior design shop, which is inaccurate, but it doesn't matter because a cage of scaffolding obstructs the words. Adjacent signage is in Chinese. Even the address is a misdirect, the number affixed to a door leading to upstairs apartments. If you weren't looking for this place, your eye would skate right past it.

But if you have an appointment and can figure out that address-number brainteaser, you might notice a scrap of writing on a piece of paper taped into the window at about waist level. It says BOOKER AND DAX.

A savvy New Yorker would know that Booker and Dax is the name of a homey, brick-walled bar on the Lower East Side, about twenty blocks north of here. Drinkers revere the place — it is, arguably, one of the most scientific drinking establishments in the world. Cocktails at Booker and Dax aren't poured so much as engineered, clarified with specialized enzymes and assembled from lab equipment, remixed from classic recipes to more exacting standards by a booze sorcerer named Dave Arnold.

The Chinatown storefront is the sorcerer's workshop.

Trained as a sculptor at Columbia University, former director of culinary technology at the French Culinary Institute, technologist behind some of the world's most experimental chefs, host of a popular radio show and blog on cooking techniques, Arnold is more than anything an inventor — of gadgets and devices, yes, but also of cocktails. He makes familiar drinks taste better than you'd believe, and crazy drinks that taste fantastic.

Stocky, with spiky salt-and-pepper hair, Arnold is talking from the instant he comes through the door. He squirts himself a glass of sparkling water, carbonated via the workshop's built-in CO_2 line to his exact specifications — he likes bubbles of a particular size — and starts running through a bunch of projects. The sorcerer is in.

The workshop is narrow, maybe twenty feet wide, and the basement is wired for 220 volts and full of power tools. On the main level, a whiteboard covered in project notes and a drying rack for laboratory glassware dominate one wall. The other is all shelves, books on the right and then bottles of booze. Arnold recycles bottles to hold whatever he's working on; ribbons of blue tape affixed over the original label say what's *really* in them. For example, a square-shouldered Beefeater gin bottle is half-full of brown liquid instead of clear, a dissonant image for anyone who has spent significant time staring at the back shelves of bars. Arnold pulls the bottle down and puts it in front of me, alongside a cordial glass. "Only take a little," he says. The handwritten label reads "25% cedar." I pour a half-ounce and take a quarter-ounce sip. It tastes like stewed roof shingle. Arnold watches my face crumble inward, and then snorts a little. He hasn't quite got that one right.

Further to the left, after the bottles, are white plastic tubs and bottles of chemicals. "I don't even know what some of this is," Arnold says. He pulls a tub off the shelf and reads the label. "What the hell is 'Keltrol Advance Performance'?"

Xanthan gum is what it is — an emulsifier, good at making combinations of liquids and solids stick together and stay creamy. In fact, most of Arnold's chemicals come from one of three classes — thickeners like the Keltrol, enzymes to break down proteins, and fining agents, things to help pull solid ingredients out of liquids. "My standard response to a new fruit or flavor is to clarify and see what happens," Arnold says. Gelatin and isinglass are good for removing tannins; chitosan (made of crustacean shells) and silica can pull solids out of milk. But vegans don't eat chitosan, gelatin, or isinglass — they're all animal products.

Arnold would like another option to offer at the bar. Chitosan made from fungal cell walls might get past the vegan barrier but doesn't clarify as well, he says, and neither does the mineral bentonite. Arnold also uses agar sometimes; it comes from seaweed. "I prefer agar clarification to gelatin," he says. "There's a flavor difference. Sometimes it's a benefit and sometimes it's a detriment. Depends on the application."

The point of all this stuff is to bring to bear the most sophisticated chemistry and lab techniques in the service of one singular, perfect moment: the moment when a bartender places a drink in front of a customer and the customer takes a sip.

So, for example, Booker and Dax makes a drink called an Aviator, a riff on a classic pre-Prohibition cocktail, the Aviation—that's gin, lemon, maraschino liqueur, and a bit of crème de violette. Made properly, it has a kind of opalescent, light blue hue and an icy citrus prickle. Arnold's version uses clarified grapefruit and lime, and it actually manages to improve on the original in terms of intense, gin-botanical-plus-citrus flavors while remaining water-clear. Alcoholic beverages are, in their way, much more complicated than even the most *haute* of *cuisines.* This is the kind of insight that drives Booker and Dax. Though Arnold doesn't really cop to that. "I'm not trying to change the way people drink. I'm trying to change the way we make drinks," he says. "I'm not trying to push the customers out of their comfort zone."

Quite the opposite, in fact. Arnold says that all his tinkering and tuning, all the rotary-evaporator distillation and chitosan fining, is about pushing people *into* a comfort zone. He's trying to take a rigorous, scientific approach to creating a perfect drinking moment, every time.

That said, while appreciating Arnold's sorcery doesn't require that a customer know the secret to the trick, it helps if the customer at least notices the magic. "Sometimes," Arnold acknowledges, "if a customer doesn't know anything about what we're doing, it can be problematic." In the early days of Booker and Dax, when Arnold was still working behind the bar every night, a guy came in and ordered a vodka and soda. It's arguably the dumbest mixed drink ever invented. In most bars, the bartender fills a tumbler with ice, pours in a shot of cheap vodka—not from the shelves behind the bar but from the "well" beneath it, where the more frequently used house labels are—and then squirts in halfheartedly carbonated water from a plastic gun mounted next to the cash register.

Not at Booker and Dax, though. Arnold thought about it for a moment and told the guy he could make one, but it would take ten minutes, and could the customer please specify exactly how stiff he wanted it? Arnold was going to calculate the dilution factor that would ordinarily come from ice and soda, titrate vodka and maybe a little clarified lime with still water, and then carbonate the whole thing with the bar CO_2 line.

It seems like a lot of trouble in the service of an unappreciative palate. "Why serve it at all?" I ask. "Vodka and soda is a crap drink."

"I think a vodka and soda is a crap drink because it's poorly carbonated," Arnold answers. "If I can make it to the level of carbonation I like, it won't be crap. I will not serve a cocktail that will make me sad."

I push the point. "But the customer wants a crummy vodka and soda, with soda from a gun, because that's what he's used to."

"Look, it's not our place to judge people's taste preferences. But I won't serve you crap." Arnold pauses for a moment, sips at his house-carbonated water. "I've never had someone not like the better version."

I've had perfect bar moments. They led to this book. Here's one: I was supposed to meet a friend for an after-work drink on a swamp-sticky Washington, D.C., summer day, and I was late. I rushed across town to get to the bar and showed up a mess, the armpits of my shirt wet, hair stuck to my forehead.

The bar, though, was cool and dry—not just air-conditioner cool, but cool like they were piping in an evening from late autumn. The sun hadn't set, but inside, the dark wood paneling managed to evoke 10 P.M. In a good bar, it is always 10 P.M.

I asked for a beer; I don't remember which one. The bartender nodded, and time slowed down. He put a square napkin in front of me, grabbed a pint glass, and went to the taps. He pulled a lever, and beer streamed out of a spigot. The bartender put the glass of beer in front of me, its sides frosting with condensation. I grabbed it, felt the cold in my hand, felt its weight as I lifted it. I took a sip.

Time stopped. The world pivoted. It seems like a small transaction — a guy walks into a bar, right? — but it is the fulcrum on which this book rests, and it is the single most important event in human history. It happens thousands of times a day around the world, maybe millions, yet it is the culmination of human achievement, of human science and apprehension of the natural and technical world. Some archaeologists and anthropologists have argued that the production of beer induced human beings to settle down and develop permanent agriculture — to literally put down roots and cultivate grains instead of roam nomadically. The manufacture of alcohol was, arguably, the social and economic revolution that allowed *Homo sapiens* to become civilized human beings. It's the apotheosis of human life on earth. It's a miracle.

Two miracles, actually. It took 200 million years of evolution to make the first one happen. Fermentation, the process by which a fungus we call yeast turns simple sugars into carbon dioxide and ethanol, is a breathtakingly complex bit of nanotechnology. Fermentation and ethanol happened on earth long before we humans got here, and ethanol's pleasant effects on our brain are a mere side effect of its use as a chemical weapon in the invisible, eternal war among the microbes with whom we share our planet.

Despite its importance in all sorts of industrial chemistry, the biochemistry of fermentation is still fodder for research. It wasn't so long ago, after all, that the greatest chemists and biologists in the world were arguing about what yeast was. Figuring out that *Saccharomyces cerevisiae* — brewer's yeast — was alive and was the thing that did the fermenting made Louis Pasteur famous and gave rise to modern cell biology. The genetics of the present-day versions of fungus still have secrets to tell: when it developed the ability to make ethanol, and why, and when we humans tamed it to our own ends.

It wasn't until about 10,000 years ago that we humans took control of fermentation for ourselves, entering into a partnership with that fungus long before we knew what it was. We domesticated that microbe, the same way we domesticated dogs and cattle, to do a job: make drinks.

Two thousand years ago, give or take, we humans built the second miracle for ourselves: distillation, one of the earliest tools used by the earliest scientists. Invented by alchemists searching for the fundamental spirits that inhabit everything on earth, the still accidentally gave rise to an entirely new way to convey flavor and aroma, and an array of drinks that became a staple of human consumption. Plus it gave rise to the modern study of chemistry and made possible our petroleum-based economy.

Those miracles make the bar moment possible; what happens in the seconds and hours after that first sip, or second cocktail, is no less amazing. Ethanol has a flavor unlike anything else, and it conveys other flavors unlike anything else. Making it is a craft — the people at Wild Turkey or Abita or E. & J. Gallo don't have to understand molecular biology, yeast enzyme kinematics, metallurgy, or the organic chemistry of polycyclic aromatic hydrocarbons. (Though they often do understand those things.) They know that the shape of a still, and the metal it's made of, changes how a spirit tastes, and they know that different wood in the aging barrels alters the flavor of the final product. (Japanese oak makes whisky taste spicier than the American oak used in bourbon and Scotch. Weird, right?)

People sometimes think science is about discovery. But the action in science, the fun part of doing it (or reading about it), isn't answers. It's questions, the stuff we don't know. Behind every step of the process that produces fermented beverages and then distills those into spirits, there is deep science, with a lot of researchers trying to figure it all out.

That's what this book is about. The bar moment is the culmination of the human relationship with our environment, the pinnacle of our technology, and a critical point for understanding our own bodies, brains, and behavior. William Faulkner is supposed to have said, "Civilization begins with distillation," but I'd push even farther — beyond just distilled spirits to wine, beer, mead, sake . . . all of it. Booze is civilization in a glass.

Out of a shared love for film noir and Los Angeles history, my mom and I once went to dinner at Musso & Frank Grill on Hollywood Boulevard, one of the oldest restaurants in the city — it traces its origins to the 1910s. My parents were mostly wine drinkers when I was a kid, but Mom inherited an occasional hankering for a Martini from her mother, so with her steak she ordered one — rocks, two olives. (Gin. Duh.)

The waiter wouldn't bring it to her. The ice ruins the drink, he said. Mom got her Martini up — shaken or stirred with ice and then strained into a cocktail glass. And I learned an important lesson: drinking has rules. The drinks are better one way than another. Behind the behaviors and preferences of the bar are algorithms. And algorithms . . . well, people can figure out algorithms.

When I was in graduate school I was poor, but every so often I'd scrape together a few bucks and go to a fancy restaurant in downtown Boston for dessert. The bar was, for the time, unusually well-stocked in single malt Scotch, so when my father and his credit card came to visit, I took him there and told him we were going to drink some. Neither of us had ever had it.

It was a weeknight, not too busy, and the bartender was happy to show off. Dad and I each picked different labels at random and asked the best way to drink them. Neat, with a water back, was the answer. So that's what we got. Our glasses of whisky arrived, and we both took our first sniff and sip. And, at the same time, said, "Oh, crap." Because we knew it was going to become an expensive hobby.

It did. In fact, a few years later I told Dad that I'd read enough. I was going to Scotland for a week to visit distilleries. He told me he wanted to come. "OK," I said, "but this is a distillery trip. I'm not going to museums or castles." He agreed, and never once suggested that he wanted to play a round of golf while we were there — though we did end up going to a castle. What I really wanted to do was drive down to Campbeltown, in the far southwest, because that's where they make the best whisky in the world. A century ago the place had dozens of distilleries; now, only a handful. One, Springbank, is exceptional.

Like all producers of single malt whisky, Springbank ferments its own mash—basically beer—and distills it. But it's also one of the last distilleries in Scotland that malts its own barley, warehouses its own casks, and bottles its own product on the premises. It's a trifecta of craftsmanship. The distillery itself is a gray place behind high walls in the old part of town, with three bright copper stills the size of houses, one slightly different in shape from the others—the shape of the still has a major impact on the final flavor of the spirit.

At eighteen years old, the oldest bottle the distillery sells, Springbank tastes like honey and vanilla and tobacco and lemon peel and leather. They used to sell a twenty-five-year-old bottling. At that age, the leather mellows out. Today, a bottle will run you upwards of $600. It was cheaper when we drank it at my wedding.

Now, that description of what the whisky tastes like—while true—makes Springbank sound the way the distilling world would like their products to sound. I didn't get that description from a book or a label, but now that you've read it, you'll taste all those things in the drink, too. Suggestion is a powerful force when it comes to booze, especially at a cost level the industry calls "super-premium." When you pay that much for something, you want it to taste special.

This is marketing, and it only has a little to do with what actually goes into a bottle. Single malt whisky like Springbank is artisanal, crafty, a creature of hundreds of years of tradition and experience. Old Scotsmen ladle samples out of aging, bull-sized casks and then smell them with noses so talented they can tell if, yes, this barrel can lay up another decade, but this one over here is ready for old women to bottle by hand. Whisky marketing—and in fact the marketing for most booze—grabs hold of, ties up, and forces those kinds of traditions to dance for money. The biggest corporations in the world will talk your ear off about how something that they measure in millions of gallons per year comes from a recipe handed down from generation to generation, is produced on ancient stills in the Highlands, and would you care for a wee dram, laddie?

In its rush for historicity, that story ignores or elides what's really important about booze — the thing that actually attracted me to it in the first place. Yes, the drinking and making of the stuff has its pleasures. But another kind of connoisseurship takes control of the story of booze away from the marketers and gives it back to the makers and the drinkers. It starts with a simple question: "How did they make that?"

A lot of people drink. According to the Centers for Disease Control and Prevention, 65 percent of Americans over eighteen say they had at least one drink in the previous year. In 1999, alcoholic beverages had $38 billion in revenues; by 2010, that figure was up to $58 billion. In 2011, Americans consumed more than 465 million gallons of distilled spirits, 836 million gallons of wine, and 6.3 billion gallons of beer. A single serving of beer or bourbon with a mixer has about 125 calories, which means that a committed social drinker might get as much as 10 percent of their daily caloric input from ethanol. But few people, even the drinkers, really know what alcohol is, where it comes from, why it tastes the way it does, or what it actually does to them. It's a mystery — to drinkers, maybe, and the marketers don't really care. But behind the walls of wineries and breweries, inside distilleries, and at research laboratories around the world, the mysteries are getting solved. Rather than letting the marketers tell us what to think about the alcohol we drink, science gives us a tool to understand it on our own terms.

Most big cities now have bars that pride themselves on fresh ingredients and deep dives into the back catalog of cocktail recipes — or innovative new recipes their bartenders have come up with themselves. Historians team up with forensic chemists to re-create obscure ingredients for pre–Prohibition era cocktails, and stores as prosaic as BevMo! stock them for thirsty big-box-store shoppers. Transnational beer companies buy microbreweries, or make their own versions of small-batch beer. Alcohol can actually be a hobby, if you have the money, inclination, and intestinal fortitude. From the perspective of my wallet, I have the booze industry exactly where it wants me.

Connoisseurship overlaps with my geekier tendencies: If you love something, my theory is, you're supposed to ask what makes that thing tick. It's not enough to admire the pretty bottles filled with varicolored liquids behind the bar. You're supposed to ask questions about them — what they are and why they're different, and how people make them. The only people who can get away with going that far down a rabbit hole are journalists, scientists, and three-year-olds. And three-year-olds aren't allowed in bars.

The chapters to come follow a sip of booze on a birth-to-death journey via your tummy. We're going to start with the life of yeast, the microbe that makes alcohol and helped spawn the fields of cell biology and organic chemistry. Then we'll talk about sugar, which is what yeast eats — and, I'll argue, the most important molecule in the universe. When we talk about sugar, we're talking about agronomy and the human relationship with plants, about how we chose what to leave to nature and what to domesticate and make our own. Sugar will also give me an excuse to detour into one of the unloved and unknown microorganisms that are just as critical to boozemaking as yeast. My favorite — a fungus called koji — might have become far more important, if it hadn't been for one or two turns of fate.

Figuring out yeast and sugar will let us move to fermentation, which is fundamentally the biology of how yeast eats sugar and makes alcohol. But it's also one of the earliest examples of how human beings started taking natural phenomena and bending them to our own needs.

Distillation, the next stage, is an even better example of human ingenuity. It applies the products of fermentation and takes technology and engineering to make them into something both less and greater than they were before. The invention of distillation happened at just about the moment when human beings started using technology to improve their lives, and that was no accident. It started with alchemists in ancient Egypt, and from there the new tech took side trips into the development of medicine, physics, and metallurgy.

Between the time someone makes your booze and you drink it, that liquid often spends time in a wood barrel — what people in the game call "maturation." It's a whole other set of chemistry, as much about the basic components of wood as the liquid stored inside it. Aging is also an economic chokepoint for people who make alcohol — and they've tried a bunch of science, some good and some shaky, to speed up time as applied to drinks before sale.

That'll put us at the pivot point, the bar moment I've been talking about. What comes after is the transition from external to internal. First we'll deal with the strange science of how human beings taste alcoholic beverages, a field where neuroscience and psychology bump up against each other's limitations. The hundreds of molecules that give a distilled spirit its flavor haven't yet been fully characterized. Peat, the partially decomposed mix of sphagnum moss and other plants that gives Scotch a smoky, earthy flavor, can have a different chemical composition depending on its region of origin — a biomolecular basis for what French vintners call *terroir* in wine. And in 2010, chemists at the University of Cincinnati (working, of course, with physicists at Moscow State University) discovered that flavor differences among the purest vodkas — composed of nothing but ethanol and water — are due to differences in the strength of the hydrogen bonds between the two ingredients. How we taste and smell the juniper in the Dutch-style gin that Anchor Distilling makes (to pull one example from the blue) is biology and genetics so complicated that it won the people who figured it out a Nobel Prize.

Figuring out what alcohol does to the human body and brain requires untangling even more difficult neurobiology — but drops a shot or two of sociology and anthropology into the cocktail, too. An example: we know people get drunk, and some people get addicted, but despite a century of research, nobody knows why. For that matter, nobody really knows why getting drunk feels the way it does.

In the end (as a digestif, if you will) I'll look at what happens when you have much, much more to drink than just a sip. The science on hangovers is far paltrier than you'd expect for something that affects so many people and feels so awful. In fact, it wasn't until the past few years that researchers even agreed on a working definition of a hangover, much less started trying to figure out its causes (and cures) in earnest. But a few brave researchers (and even braver research subjects) are finally working on it. It turns out everything you learned about hangovers in college is wrong.

No matter how much the makers of booze and the people who study it understand, there's more that they don't. The field still has mysteries. And that's awesome. Booze lives at a persistent point of conflict in the sciences — that place where subjective experience bumps into objective evidence. Researchers have turned their analytical gear on fermentation and distillation, and learned a lot, but in some cases, they still haven't answered basic questions. Ethanol is one of the few legal drugs of abuse — and the only one that nobody really understands at the functional level. Yet entire businesses — not to mention vast swaths of popular culture — are built around describing the flavors of those drinks and getting people to choose which ones to pay money for.

Scientists would love to get those two parallel lines to intersect. They'd like a list of molecules that, when found in the right quantities in a drink, made that drink taste better (and maybe sell better) than drinks without them. They'd like a replicable explanation for drunkenness in the brain that correlates to the behavior of people acting under the influence of ethanol. But they haven't made any of that happen yet.

This book is not a textbook. You can find one if you want; academia is full of people studying yeast, beer, wine, and spirits. You won't get instructions for building a still or making your own mead here, and neither will you get many recipes for cocktails. (I put in a few of my favorites.) And one point of style: People argue about whether to spell the name of the drink distilled from grain as *whisky,* as they do in Scotland, or *whiskey,* as they do in Canada and the United States. I'm going to drop the *e*. Deal with it.

The road from yeast to hangovers is a story of 10,000 years of obsessive work perfecting an ingredient at the center of human ritual and recreation for as long as we have been civilized. But behind that story is a less obvious one, a tale of the fundamental cleverness of our species.

We humans sometimes face forces we don't understand, and occasionally we take control of them and create technology with them.

Understanding our relationship with alcohol is about understanding our relationship with everything — with the chemistry of the universe around us, with our own biology, with our cultural norms, and with each other. The story of booze is one of intricate research and lucky discoveries that shape, and are shaped by, one of our most universal shared experiences. The human relationship with alcohol is a hologram for our relationship with the natural world, the world that made us and the world we made.

One

YEAST

A COMMERCIAL BREWERY IS really a factory. Raw ingredients like grain and water go in one end, flow through pipes and tanks, and beer comes out the other side. But you could gut and replace all those pipes and tanks, switch from one grain supplier to another, swap out the walls and the controllers, and the same beer would still flow from the taps, metaphorically speaking.

The one thing the brewery cannot afford to lose is a finicky microbe that is the not-so-secret power behind the whole show. If you are a brewer and you plan to make a product people like, and keep making it the same way, you must maintain your yeast. The same goes for wineries, and even for distilleries — before you can distill a spirit, you have to have something fermented to start with. If you lose your yeast, you're dead.

In fact, "We're dead" is exactly what went through Rebecca Adams's mind when she arrived at work one day in late November 2009. Head of the lab at Jennings Brewery in England's Lake District, Adams had slogged her way to work after a massive flood — sixteen inches of rain in twenty-four hours that pushed the Rivers Cocker and Derwent over their banks and put the stone walls, arched bridges, and whitewashed buildings of the medieval town of Cockermouth, at the confluence of the two, under ten feet of water. When Adams got to Jennings, she saw that they'd lost most of their service machinery—the boiler, the air compressors, the chillers. But that wasn't the worst part. Jennings makes real ale, an endangered species of drink made almost nowhere else in the world. Technically, real ale requires a specific variety of yeast that, during fermentation, floats on top of the sugary wort instead of sinking to the bottom. Ale is a rich, big-bodied, chewy experience very different from German-originated lager- and pilsner-type beers. For the British, ale is a cultural touchstone. The yeast is one of the things that make ale special. At Jennings after the flood, the yeast was gone. Drowned.

"I was there by half past six together with most of the operators," Adams says, "and we honestly didn't know whether we'd be working again." They could replace the machines. The yeast was a whole other kind of problem.

The miracle of yeast is awesome enough to strain credulity. It's a fungus, a naturally occurring nanotechnological machine that converts sugar to the alcohol we drink. It breeds pretty much everywhere and is one of the organisms on which scientists have built much of our knowledge of how life works . . . and, postscript, it also makes possible the baking of bread. Unbelievably practical and staggeringly improbable—yeast is practically science fictional. When Douglas Adams invented an equally useful biotechnology, the language-translating Babel fish, in his *Hitchhiker's Guide to the Galaxy*, he included an aside that said it was so fantastic that it disproved the existence of God (because it would imply the existence of a benevolent Creator, and proof belies faith, but without faith there can be no God. Poof! God "vanishes in a puff of logic"). And yet: here's yeast. Eats sugar, makes ethanol.

I'm not saying yeasts are divine . . . but exactly 200 years before Douglas Adams published his book, Ben Franklin told essentially the same joke, only more succinctly. Franklin said that rain falling on grapes, which could then be turned to wine, was "a constant proof that God loves us, and loves to see us happy." Granted, Franklin didn't know he was talking about yeast, because nobody knew what yeast was until about 150 years ago. Yet we humans became dependent on it, without even knowing it was there. In utter ignorance, we made yeast into our partner. Absent knowledge of the mechanisms that made yeast work, the stuff was a miracle — the provenance of people who might as well have been wizards. Divining its secrets, figuring out that a living creature invisible to the naked eye was an agent of transformation, sparked a scientific revolution.

Yeast lives as a single-celled organism that is neither plant nor animal, neither bacteria nor virus. The fungus family includes every mushroom you've ever seen, lichens, rusts and smuts, athlete's foot and the *Candida* that infects people's most intimate parts, Dutch elm disease, the parasite that causes dandruff, and slime mold, the single largest creature on earth. Like animals, fungi tuck their genetic material — their DNA — into a structure inside their cells called a nucleus. Like plants, fungi have walls around their cells that provide strength and protection. In plants, that wall is mostly cellulose and lignin — the hard-to-break stuff you probably know better as "wood." Fungi mix in a little chitin, which is almost identical to cellulose save for the addition of nitrogen — and it's the main ingredient in insect exoskeletons and octopus beaks. Nature, weird in tooth and claw.

Yeast was the first eukaryote — that is, the first creature with cells and nuclei — to have its genome sequenced. That was in 1996; biologists were in a rush to see what its DNA looked like because in a way yeasts are the fundamental unit of cell biology. Yeasts grow quickly and easily in a lab, but because they have nuclei just like we do, they are an excellent model system for life like us. They are the critter that has taught us much of what we know about the cellular world, making them, as one article puts it, "famous and atypical . . . an excellent model for the basic features of eukaryotes and for experimentation, but a poor model for other fungi."

All that, and they ferment. Assuming that combustion — fire — is civilization's most important chemical reaction, then yeast are responsible for the chemistry in the number-two slot.

After the flood in Cockermouth, Jennings Brewery started to clean up. The company bought new equipment, but the most important thing the brewers needed was in a steel tank full of liquid nitrogen, waiting for just such an emergency in a four-story building in the town of Norwich, 290 miles to the southwest. That's the National Collection of Yeast Cultures, a research lab with a side business in preserving copies of yeast strains used by British brewers — offsite backups kept in case of something like, oh, let's say, a massive flood.

Jennings is one of several breweries owned by the same company; once the directors decided that they'd reopen it, they knew they had to keep production going. "You can't afford to lose your presence on the bar. You need to still have the Jennings beers out there," Adams says. "So the other breweries down south brewed the beer under our name using our recipe." They called the NCYC and got a sample of the Jennings yeast, packaged in a form called a "slope," suspended in agar gel inside a glass vial. The other brewery grew it up into a quantity large enough to brew with, ordered the right recipe of barley and hops, and "they were able to brew Jennings beer with Jennings yeast," Adams says.

In February of 2010, Jennings reopened with all-new service equipment — most of it relocated to higher floors in case of another flood. But for Rebecca Adams, it was the return of the yeast that signaled the true reopening of the brewery. "Once we were working again, they sent a five-barrel tank of our yeast back up to us, which was an exciting day," she says. "It felt like we were going to have a future again."

The Institute of Food Research — parent organization to the NCYC — used to employ 2,000 people. Today just 100 remain, and on the day I visit a church-like quiet extends all the way down a hall to a blue door, where the curator of the collection, Ian Roberts, has an office behind a lab now used as a robotic yeast-handling facility. Roberts himself looks a bit like a lighter-haired John Kerry.

"The NCYC grew out of a brewing collection," Roberts says. "I suspect most of the collection originated with British ale brewing." He now has 4,000 different samples, and of those up to 800 are brewing yeasts. (The rest are random wild types and the kind of yeasts that spoil food and infect people with malfunctioning immune systems.) In the 1920s, a brewers' industry group maintained the collection; in 1948, it got nationalized. "We will provide services to brewers, pharmaceutical companies, the general public," Roberts says. "Pretty much anybody who needs yeast."

Down the hall, another blue door opens onto a narrow room with mint-green walls. Behind a low chain and a chest-high tank of supercold liquid nitrogen is a squat cabinet the shape and size of a washing machine, with a round door on top. It looks like what it is: a deep freeze. "That is the National Collection of Yeast Cultures," Roberts says. In addition to the thousands of research species and strains are another 650 or so in the R Collection, the brewers' reserve — among them, the Jennings sample. "Our role is biodiversity preservation," Roberts says. "Because we know what market forces do to biodiversity. There's over 100 years of microbiology in there." To keep track of all the different yeasts, Roberts's team uses three-by-five index cards and a database on an old Macintosh clone.

The samples inside the freezer live in half-inch-long pieces of red drinking straw, sealed at both ends and then put into tiny screw-top vials called freezer tubes. They're frozen because one thing yeasts do really well is mutate. They share genes with neighbors and generally fail to remain stable. Go twenty generations down the line and you have a different strain from the one you started with, which is bad news if you're trying to make the same beer, batch after batch. Done right, freezing preserves the yeast indefinitely — ready to be "grown up" and sent to users like the Jennings brewers.

The NCYC has its own backup collection as well, in the form of freeze-dried yeast powder sealed inside glass ampules. It's upstairs from the freezer, behind a six-inch-thick wood and metal door with a big locking handle. It's cold inside — Roberts closes the door, but doesn't pull the handle shut. He knows it can't be locked from the inside but he's overcautious by nature.

In an inner room, behind another refrigerator door, are what look like filing cabinets. Roberts pulls a drawer open to reveal bowls filled with labeled, two-inch-long sealed ampules. Their tops have the pulled-taffy look of glass that's been melted over a Bunsen burner. Inside each is a bit of cotton and a puff of white dust. That's the yeast.

"We've got very old ampules," says Chris Bond, the collection manager. For twenty-four years, Bond has been the guy who responds to queries from people who want samples — spoilage yeasts, stuff to test preservatives, or strains to make beer. "Small brewers, microbreweries, come to us to look through the collection and re-create beer from history, like the 1940s," says Roberts. Big shots like Watneys were once based in Norwich; the town, a medieval walled city, had more than 300 pubs. Some of those defunct brewers left their yeast in the collection. And some brewers try to go back in time even further. "We actually had someone trying to recreate a South American beer from the Incas," Roberts says.

One thing about Bond's lab: it doesn't smell. Every lab I've been in that studied yeast smelled like some funkier version of baking bread, maybe less sweet. The air in Bond's lab was clear. That's because they're not doing fermentation, the researchers told me. And they're not incubating the yeast — growing it up. They're just preserving it. "When you decant it, it smells like beer, because we use brewer's wort," Bond says. Other than that? Nothing.

The R Collection, the safety-deposit box service, isn't for research. For around £250 a year, a company gets to keep a copy of their yeast strain at NCYC. And some brewers, it must be said, aren't as convinced as the Jennings folks the strain is vital. That schism runs right through the alcohol business. Roberts recalls a brewers' trade meeting where "you could split them fifty-fifty, the ones that thought yeast was of no consequence, just another chemical, and the ones who thought their yeast was the be-all-end-all." Roberts says his job is just to put yeasts into his archive and leave them alone (and revive them when needed). Most of the time, he and his team don't even get tastes of the beers they help create — a fact about which he is, you should excuse the expression, bitter.

As we walk back into the conference room for a lunch of charmingly British sandwiches — cheese and Branston pickle, ham and mustard, egg salad, prawn — Roberts hesitates. "I don't know if you touched anything in the lab," he says to me, "but you might want to go wash your hands. Yeast are pretty harmless, but . . ." He trails off with a little shrug and a smile. I go straight into the bathroom, head for a sink, and put my hands under water as hot as I can tolerate.

An archive like the NCYC might seem routine, but that's only because people like Roberts know what they're doing. Go back 2,500 years and things were different. When the philosopher-scientist Aristotle wondered why sugary liquids could become alcoholic over time, he guessed it had to do with a *vis viva,* a vital force that animated all living things toward some goal. Grape juice *wanted* to mature into wine, and its eventual decay into vinegar was akin to death. Even as recently as 1516, the German law governing beer, the *Reinheitsgebot* — the world's first food safety law — mandated that the only ingredients be barley, water, and hops. No yeast, because no one knew what it was. The Bavarian dukes who passed the law didn't even know they had a mystery on their hands.

In any successful fermentation — that is, any time juice or honey or whatever ended up yummy and alcoholic instead of spoiled and sour — a kind of haze would condense out of the liquid, and that sediment would help the next fermentation work, too. People called that sediment "yeast" because of its action. The French and German names (*levure* and *Hefe,* respectively) derive etymologically from roots meaning "to lift," as in rising bread. The English "yeast" comes, via the Dutch "gist," from the Greek word for boiling. Getting the gist of something is literally boiling it down.

Some of history's most important researchers, the people who founded modern science, dedicated at least part of their lives to studying fermentation. A century and a half after the *Reinheitsgebot,* Anton van Leeuwenhoek, the inventor of the microscope, put a drop of fermenting beer under his brand-new lenses and saw individual yeast cells. He sent his drawings and descriptions of these oval and spherical bodies to the Royal Society of London, but nobody could figure out what they were, and they stopped paying attention. That lack of interest lasted a century and a half.

Finally, in 1789, someone picked up the work — albeit from a different angle. Antoine Lavoisier, discoverer of oxygen and hydrogen, published the first quantitative work on the conversion of sugar to ethanol and carbon dioxide. Lavoisier — perhaps because, as some writers have suggested, he worked at a big tax firm — was really good at chemical accounting. He figured out that at the end of any chemical reaction you should have the same amount of stuff you started with. This was the origin of the law of conservation of mass, the one that says matter is neither created nor destroyed, but can be changed.

When Lavoisier learned that grape juice was up to 25 percent sugar, he hypothesized that sugar was what turned into ethanol — somehow. And he did a slick experiment to prove it. Lavoisier fermented pure sugar. Then he separately burned the same amount of sugar and the resulting ethanol and, using careful scales he built, measured the products. He started with 26.8 pounds of carbon (in the sugar) and ended with 27.2 pounds of carbon (in ethanol and carbon dioxide). That's basically equal, within the bounds of experimental error. But since the yeast was pretty much the same by weight and on appearance, Lavoisier ignored it.

Famed chemist Joseph Louis Gay-Lussac brought more precision to the measurements and the equation, but even he threw out the yeast. The French Revolutionary government cut Lavoisier's head off in 1794, and nine years later, in 1803, the Institut de France offered a gold medal weighing a kilogram to anyone who could figure out how fermentation worked. Nobody won.

Twenty years later, the French wine industry alone was worth £22.5 million — in *1820s* pounds. That's $2.5 billion today, and doesn't even count beer, cider, or anything distilled.

Finally, in 1837, a German physiologist named Theodor Schwann took on the problem. He suggested that those microbes Van Leeuwenhoek had seen were the agents of fermentation. Schwann was a cell-biology ace who'd go on to characterize cells critical to the nervous system; today we call them Schwann cells. He was the first person to figure out that yeasts could reproduce asexually, that they eat sugar, that they need nitrogen to survive, and that they excrete ethanol. He called them *Zuckerpilz,* and a colleague named Franz Meyen riffed on that to rename the genus. *Zuckerpilz* is German for "sugar fungus"; Meyen translated that into Latin and got *Saccharomyces.* The kind used for brewing and baking became *Saccharomyces cerevisiae,* as in *cerveza* — beer.

Problem solved, right? Give those guys a kilogram of gold. Only . . . this is when the chemists get involved.

Chemists and biologists have always been at odds; chemists think that they're explaining at a more granular level of detail what biologists purport to study more holistically. (And physicists? Don't get them started.) Two German chemists named Friedrich Wöhler and Justus von Liebig and their teacher, a Swede named Jöns Jacob Berzelius, stepped in to vigorously denounce the idea that some kind of microbe could make alcohol. The chemists, perhaps predictably, believed that fermentation was a chemical process — something that just happened on its own when you left fruit juices alone. No mythical microbes were required.

These guys weren't just shmucks off the street. They had a reputation for being right about stuff. According to some histories, Berzelius is the scientist who first called molecules that contained only carbon, hydrogen, oxygen, and nitrogen "organic," because they're found in living things — thus inventing "organic chemistry," the college course that destroys the hopes of so many kids who think they're going to be doctors. Von Liebig, for his part, came up with the idea of having chemistry students study in actual laboratories. And the three men were together responsible for the discovery of chemical isomers, substances that share the same elemental ingredients but have different properties, owing to different structures. Basically, you put the same ingredients together in a different order, and instead of getting a cake you get a sausage.

Also, they were funny. When Wöhler accidentally synthesized urea, one of the main constituents of urine, he famously wrote to Berzelius,

"I can no longer, so to speak, hold my chemical water and must tell you that I can make urea without needing a kidney."

Von Liebig thought that when yeast died and decomposed, it gave off some kind of vibration that broke apart sugar, which then rearranged itself into ethanol. In hindsight that sounds a little nuts, but the idea wasn't crazier than anything else floating around at the time. So when Schwann, this upstart biologist, had the temerity to disagree, von Liebig and Wöhler directed their brains, scientific and satirical, at undermining him. In the scientific journal they ran, *Annalen der Chemie und Pharmacie,* they wrote (anonymously) a comical description of yeast ("the shape of a Beindorf distilling flask") fermenting, seen under a microscope: "In short, these infusoria eat sugar, eliminate alcohol from the intestinal tract and CO_2 from the urinary organs." A magical tiny creature that pees carbon dioxide and poops out beer? Please.

What we know now makes the chemists look misguided, but they had a point. Early microscopists weren't working with the best gear, and they had a reputation for seeing things that weren't really there. The argument went on well into the 1850s, which was a big pain for the world's producers of alcohol, because they had an industry based on a mystery. That didn't matter when everything was working. But when something went wrong, nobody knew enough to fix it.

As an example: From 1800 to 1815, a British blockade kept cane sugar from the West Indies and Asia out of France. So the French turned to a substitute: sugar beets, the only other plant from which you can easily extract sugar. By the mid-1850s, the Lille region in northern France was a major producer of alcohol fermented from sugar beet —profitable, but, like wine and beer, prone to problems during production.

One of the Lille-area producers, a man named Bigo, had a few vats producing ethanol just fine, but others were generating something sour, like spoiled milk. Bigo mentioned his problems to his son, and the young man told his father that one of his professors at the University of Lille might be able to help.

Bigo *fils* was talking about a dour Royalist with a reputation as a fine crystallographer, just thirty-three years old, working on isomers —those differently structured molecules with the same chemical ingredients that Berzelius discovered. The researcher's name was Louis Pasteur.

Pasteur had discovered that if you shined polarized light through some isomers—polarized light having all its wavelengths restricted to one plane, as in some sunglasses, instead of wiggling around at random—the isomer would rotate that light. These "optical isomers" were a big enough deal to get Pasteur a professorship at Lille in 1854, and the next year he started looking at isomerization in the alcohols in fermented beet sugar. Two of them, he found, were actually optical isomers of each other, a discovery that turned him on to the strange, poorly understood chemistry of fermentation.

Pasteur agreed to come have a look at Bigo's tanks. This is Pasteur before he was big. It's pre-pasteurization Pasteur, pre-germ-theory-of-disease, pre-rabies, pre-everything. When Pasteur first got interested, nobody was even sure what fermentation was. The kind that made alcohol was, back then, called "spirituous fermentation." But the production of acetic acid—vinegar—and lactic acid (in pickles and spoiled milk) looked like some kind of fermentation, too. For that matter, some researchers made a good case that putrefaction—spoilage and rot—was yet another kind of fermentation. Each process involved a mysterious transformation of *this* stuff into *that* stuff.

Now, this story's provenance isn't the best. Pasteur himself never told the anecdote about going to work in the factory of Monsieur Bigo the sugar-beet fermenter; it only came up in later hagiographies. Pasteur had actually expressed an interest in the causes of fermentation long before coming to Lille, though he himself never drank much. And his early work involved tartrates, a winemaking byproduct. So maybe he suspected that working on fermentation would lead him to living microbes. (As one of Pasteur's biographers, Gerald Geison, put it: "I believe that Pasteur began his empirical research already expecting to find a correlation between crystalline asymmetry, optical activity, and life.")

But either way, this is the legend: Upon arriving at Bigo's factory,

Pasteur saw that the off-smelling batches looked different, as if they were dirty or contaminated. So he took a few samples. Under magnification, the normally fermenting beet sugar had the same little round structures that Schwann and others had seen; the off batches had long black rods. And instead of alcohol, the spoiled batches were full of lactic acid.

Pasteur concluded that the round cells, the yeast, were somehow making ethanol. The black rods, whatever they were, made lactic acid. He backed up his hypothesis with an ingenious experiment, the kind for which Pasteur would become famous. He put sugar and nutrient minerals in flasks and added sediment from the successful fermentation to one, and sediment from the lactic fermentation to the other. The first produced ethanol; the second did not. This went way beyond Schwann's correlation. This was proof.

Pasteur didn't know how to keep the competing microorganisms out of the beet sugar, or even what they were, but he could still make a recommendation: Clean the tanks. Get rid of every last drop of the contaminated batches and start over.

Starting in 1857, Pasteur published papers and books further laying out the process. He realized that ethanol wasn't the only product of fermentation—depending on what he started with, he'd also get glycerol, succinic acid, or butyric acid. It was the first time anyone suggested that microorganisms could take in compounds from their environment, do something to them, and excrete something else. It was the beginning of the idea of metabolism, of studying the biology of how living things make energy. In 1866, Pasteur published *Études sur le vin,* his major book on wine. Another, on beer and fermentation, followed ten years later.

Von Liebig never bought it. He insisted that while yeast contained something that fermented sugar into ethanol, the yeasts themselves weren't necessary to the process; Pasteur said the yeast had to be alive for it to work. Von Liebig had a couple of good examples on his side. In 1833, French chemists purified a chemical that converted starch into sugar, no living creature required. Three years later, Schwann himself isolated a chemical he called pepsin, which could break down muscle, blood clots, and coagulated egg white. Scientists at the time called these chemicals "soluble ferments"; today we'd call them "enzymes," proteins that accelerate biological processes. (I'm going to talk about them more in the next couple chapters.) The two men fought it out in public, in journals. Pasteur kept up his end of the argument until well after von Liebig's death.

By 1897, every one of the players so far was dead and the field could finally get on with things. That year, a German chemist named Eduard Buchner paid a vacation visit to the Munich lab of his brother Hans, a microbiologist. Hans had been trying to figure out how to keep yeast extracts active for more than a few hours after cracking open the cells, and one of his approaches was to put gunk extracted from the yeast into high concentrations of sugar, up to 40 percent.

One of the preps caught Eduard's eye. It was bubbling, and he knew that meant it was fermenting. The two brothers set to work figuring out what they'd done, and eventually came up with a process for keeping the extract viable in perpetuity: They mixed brewer's yeast with sand, ground it up, added water, and squished the paste under a hydraulic press. Then they ran the press juice through a paper filter, and, as the Buchners wrote of the resulting extract, "the most interesting property of the press juice is its capacity to bring about the fermentation of added carbohydrates." They named the extract — which, though they didn't know it, was a whole bunch of purified enzymes — "zymase."

In other words, in the end Pasteur and von Liebig were both right — some substance or substances inside a living yeast cell did the actual fermentation, but yeasts are definitely the agents of the process. There's a certain elegance to the Buchners' work. For one thing, it took a biologist and a chemist working together, going against what was by then a century of animosity. Figuring out exactly what was happening in the guts of the yeast cells would take another few decades and spark a whole other field of science — biochemistry.

The realization that yeasts were behind fermentation led to a whole other set of questions. Why did some fermentations go better than others? Which yeasts were the best? In the early 1880s — when the yeast controversy was ebbing but not yet over — the man to see about microbes was Robert Koch. He was pioneering the use of a whole bunch of brand-new lab techniques, growing bacterial cultures in petri dishes and using agar as a nutrient medium. These state-of-the-art practices, now standard, allowed Koch to isolate anthrax and tuberculosis, among other astounding results. Koch would go on to codify a series of postulates still used today to verify that a given microorganism is responsible for a given disease.

In the autumn of 1882, a Danish microbiologist named Emil Christian Hansen visited Koch's laboratory. Hansen was working for Carlsberg Brewery, a purveyor of classic lager that was having problems with bitter flavors and off odors. Hansen thought he could apply Koch's ideas to the microorganisms in beer, and using Koch's techniques he was eventually able to culture four distinct yeast strains in use at the brewery. Hansen tracked the problems to one; in fact, he figured out that only "Carlsberg bottom yeast no. 1" was producing good beer. The brewery switched to using it exclusively, and in 1908 Hansen designated it *S. carlsbergensis.* (Biologists really, really care about names; in the days before genome sequencing, taxonomists argued over every little detail of a fungus's looks and behavior to figure out where it belonged. Hansen thought he'd found an entirely separate species of yeast, as opposed to merely a closely related strain, and wanted to distinguish it from the *S. cerevisiae* that floated at the top of a fermenting liquid and were more likely to make the heavier style of beer called ale.) The Carlsberg strain — or species, if you believe Hansen's taxonomy — was a lager yeast, sinking to the bottom during fermentation.

Actually, the tendency of certain yeasts to clump together and sink during fermentation — called flocculation — is still an issue for brewers and researchers today. Ale yeasts don't floc well, so they float on top of the stuff they're fermenting; yeasts used to make lager are strongly flocculent, so they stick together and sink. If you're using yeast to study, say, cancer or metabolism, flocculation is a pain in the ass. Sticky yeasts are harder to work with. But if you're trying to make a specific style of beer, or to recover your yeast after the fermentation of wine, you want to know what the floc is going on.

It turns out that the cell walls of top-fermenting yeasts repel water, which theoretically makes them more likely to stick to bubbles of carbon dioxide and float. Bottom fermenters have protein-sugar projections that stick to each other like Velcro. Shave off those hair-like fronds (they're called "fimbriae") by running the yeast through a blender, and flocculators don't floc anymore.

Bottom-fermenting lager yeasts like Hansen's *S. carlsbergensis*— now usually called *S. pastorianus,* just to make things even more complicated—have become the dominant type used by brewers around the world. Brewers and winemakers actually don't mind flocculation, because it makes it easier to remove the yeast after they've converted all the sugar. This probably explains why, after centuries of selection for the trait, brewer's yeast strains floc and wild strains of yeast tend not to. But *S. pastorianus* had no life outside industry—human beings keep it alive and thriving. Nobody knew where it came from. Nobody really knew where *any* yeast came from—what was their home in the wild? And how did humans find the yeast that would make good bread and good beer?

Those are the questions that interested a geneticist named Justin Fay. In the early 2000s, he started asking people for yeast samples, and when he moved to Washington University as a researcher, he realized he could use gene-sequencing technology to get some answers. "Despite the immense amount of information we have for *S. cerevisiae* from all of the laboratory work, we really didn't know much about where it comes from," Fay says. "Most of the samples that people had obtained were from bakers, brewers, and vineyards. So the idea at the time was that yeast was a domesticated species, sort of like dogs, or cattle, or corn." But then people started sending samples—or depositing them in living databases like the NCYC—collected outside the places where yeasts work professionally. Many came from trees or hospitals. "The question was," Fay says, "are these things that have escaped from the vineyard, like feral dogs? Or are these actually the wild ancestors?"

Fay is talking about domestication, the process of taming something wild. Actually, a better definition might be, as Fay puts it, "a species modified specifically to do something for us, for some specific purpose." This isn't just training a single animal. Domestication means breeding in traits for tameness, genetically, over generations. Cattle, for example, are domesticated—human beings eat their meat and drink their milk, and no one ever sees a wild cow. Sows on farms don't give birth to wild boars. (The difference is tusks. And rage.)

For some species, scientists have a good idea—or at least a hunch—about when domestication occurred. Genome sequencing helps with that. They can, as Fay hoped to do with yeast, look for differences between the genes of extant domesticated species and those of their cousins in the wild. Because genes mutate at a known rate over time, more differences mean a divergence farther in the past.

One classic experiment makes clear the differences between wild types and domesticated species better than any other. In 1958, Dmitry Belyaev, a biologist at the Russian Institute of Cytology and Genetics in Siberia, set out to learn how, 15,000 years earlier, wolves turned into dogs. He and his students and colleagues collected 130 undomesticated silver foxes from nearby fur farms and only bred the nicest ones—the ones that approached their handlers at feeding time instead of cowering in the back of the cages (also the ones that didn't bite). In just nine generations, Belyaev's foxes were tame as puppies. They looked like puppies, too—varicolored, with floppy, juvenile-looking ears. The foxes had taken on the visual characteristics, the phenotype, of all domesticated animals. And they were playful and liked hanging out with people.

Belyaev's experiment is still going on. As a control, the lab has mirror-universe foxes, too, intentionally undomesticated, wilder than even their wild cousins, all snarls and bared teeth. Over the years, the researchers in Siberia were able to get the same results with mink and rats, and recently geneticists have begun taking samples from the foxes to try to link the domestication phenotype to a genotype—something that's tough for any trait, but especially tough for complex behaviors.

Those experiments reflect *only* the intentional, human-run part of domestication. They don't account for the unintentional part that has to come first, when people and the organism they're working with live in wary symbiosis. But other researchers have tried. In a 2003 paper, Hungarian biologists describe an experiment similar to Belyaev's. They hand-raised wolf pups and dog pups side by side. Both grew up tame. Both grew up smart. But on tests of social cooperation, the dog pups looked to their human handlers for help, while the wolves insisted on pushing on alone. They didn't have an instinctive sense, the researchers argued, that they were in a pack-like relationship with the humans. But the dogs expected the humans to lend a hand.

When wolves first abandoned their packs for bands of hunting-and-gathering humans, learning to roll over cutely and not eat babies in return for getting a fire to sleep next to and scraps of food handed to them instead of having to hunt, were they the *Canis lupus* version of Uncle Toms? Or were they more canny? We thought we were domesticating them; maybe they were domesticating us.

You can do the same kind of experiments on microbes. In fact, it's a lot easier. Justin Fay had a living sample of wild yeast: *S. paradoxus,* a species related to brewer's yeast but not used in labs or for alcohol production. *S. paradoxus* lives on oak trees, in the bark or on "exudates," gushes of sap. Like *S. cerevisiae,* it eats sugar and excretes ethanol.

Fay and another researcher, Joseph Benavides, gathered up all the yeast samples they could find — eighty-one altogether. Most were from vineyards, but Fay and Benavides also acquired a bunch used to make sake — Japanese rice wine — and its distillate shochu. They had examples from African palm wine (made from palm sap), from Indonesian ragi (a kind of yeasty rice cake), and one from apple cider. Nineteen strains were from oak trees or the infections of immune-compromised patients in hospitals.

From those, Fay picked five genes at random and found about 180 genetic polymorphisms — places on the genome that differed among strains. When he compared them, he found that the strains most similar to *S. paradoxus* (and therefore the oldest) were the oak-exudate strains from Africa and North America and the clinical ones. Of the strains used for fermentation, the ones from Africa were the oldest, and the vineyard and sake strains had less variation than the others.

To Fay, those results suggest the following: humans domesticated *S.*

cerevisiae from African yeast about 11,900 years ago. Sake yeasts came from that same line 3,800 years ago, and vineyard strains 2,700 years ago. Fay can't be as specific as he'd like, because part of his calculation requires knowing the duration of a yeast generation — their time from birth to reproduction — and people's estimates of that vary by a factor of ten. But broadly, Fay's numbers agree with the dates archaeologists attach to the oldest examples of human winemaking and sake brewing. "I think what it really showed is that like a lot of domesticated organisms and species, there is a strong population structure for yeast," says Fay. "There are yeasts used to make wine, and those all group together. There's yeast used to make sake, and those all group together genetically. There's a genetic pattern that corresponds with what they're used for."

The strain of lager yeast Hansen isolated at Carlsberg had unknown origins. Its genotype says that it's half *S. cerevisiae,* but the other half was unidentified. In 2011, a team of Portuguese and Argentinean yeast hunters set out to solve the mystery of this mutt's parentage.

They went looking for wild yeasts in the forests of Patagonia — focusing on southern beeches, because they occupy the same ecological niche in the Southern Hemisphere as oak trees do in the north. "Our species have a fungus that infects them, called *Cyttaria,*" says Diego Libkind, a biologist who grew up in the mountainous region where the trees grow. "In spring, the presence of the fungus in the tree makes tumors. The branches get swollen." Yellow growths called galls, about the size of a golf ball, begin to grow. They're the fruiting bodies of the fungus, essentially big, spherical mushrooms, and they're almost 10 percent sugar. So various species of yeast colonize them and start chowing down. "Once they mature, they fall to the carpet," Libkind says. "When they start fermenting, you can smell the alcohol." The local indigenous people used to make a drink from the spontaneously fermenting fruiting body (which, Libkind hears, "is not so tasty"). Though the galls themselves, sold in Chile as "llao-llao," are apparently good eatin' on their own.

Eventually, the level of ethanol inside the fruiting body rises so high that only one species of yeast can tolerate it. When Libkind's colleagues sequenced its DNA, it turned out to be previously undiscovered — and its genes matched the unknown half of the lager yeast. Libkind's team named it *S. eubayanus.* They did some sequencing work on a few different genes, some of them involving sugar metabolism. "All the changes we see, we know they're the result of the brewery domestication," Libkind says. "Less efficient genes are turned off and more efficient genes are turned on." Through the simple act of switching to batches of yeast that made better beer, the Carlsberg brewers were subjecting their yeast to the same kind of selective pressure as Belyaev's team did to the foxes in Siberia. The yeast was getting, in a real sense, friendlier — easier to use, harder working, willing to make nice crisp lager in return for a scratch on the tummy.

Libkind's team isn't done. "We are going for ales now," he says. They're collecting yeast strains from around the world — distillers, sake brewers, wild fermenters — and trying to figure out what they're descended from. "Now we know that many brewing strains are a mix of different species," Libkind says. "We know many Belgian strains are actually a hybrid of *uvarum* and another *Saccharomyces* strain." The search, in other words, continues.

The really weird part is that nobody has ever found Patagonian *S. eubayanus* wild in Europe. Nobody knows how it could have gotten there to hybridize with *S. cerevisiae.* "Imported from overseas after the advent of transatlantic trade" is the only guess the researchers hazard. It's another mystery.

Presumably, early bakers, winemakers, and brewers exerted selective pressure by using particular fermentation vessels, particular vineyards, and particular regions. Whoever had the best-tasting product, or made it most reliably or most cheaply — that's who people kept buying their potables from, perpetuating the strains they used. And since yeast are notoriously prone to mutations, new strains would crop up all the time, and preserving the old ones would become a kind of sacred trust. That's why sourdough starters used to be part of young women's dowries, handed down from mother to daughter, and it's why today's boozemakers tend to return to a preserved, frozen sample of their yeast and grow more instead of just letting the same batch bubble away. It's why, before Daniel Bacardi fled Cuba in advance of the encroaching revolutionary army in 1960, he destroyed every sample of the fast-fermenting yeast strain that made his rum — Bacardi planned to start over in Puerto Rico, and didn't want the new Cuban government to have a competing product.

Even when boozemakers look like they're being casual about their yeasts, they're not. Brewers of Belgian lambic-style beer ferment in giant, open pools with whatever yeast happens to fall in (as opposed to "pitching" a purchased or carefully maintained strain). Lambics tend to be quite sour, probably because the local microbial flora include the genus *Brettanomyces* and associated bacteria that excrete acetic acid — vinegar. Usually brewers and winemakers have to follow strict sanitary procedures to keep unwanted microbes with their off flavors out of the ferment. But none of that makes lambic brewers careless about their yeast strains — one brewer, upon being told his fermenting house needed a new roof, panicked. He was sure the yeast colony that made his beer what it was lived in the rafters. So he built a new roof over the old one.

Once you know that yeast is mutable, you can use that to your advantage. Sake brewers in Japan once gauged the progress of their fermentations by the height of a big, frothy head of foam in the tank. But that meant tanks had to be much bigger than the volume of the liquid being fermented, limiting output. So in the 1960s a famous sake researcher named Hiroichi Akiyama set out to develop a new yeast. He'd seen low-foaming fermentations, and he knew that yeast clung to the bubbles, so he started doing experimental fermentations with the classic sake strain Kyokai no. 7. Akiyama would scoop off the froth, filter out what was left in the ferment — sake yeast doesn't flocculate — and breed those.

Over and over he repeated the process. He called it "the bubbling method," and eventually it yielded a new strain that didn't make as much foam. Akiyama called it Kyokai no. 701. "Today, forty years after these non-foaming yeasts were successfully selected, around 80 percent of Japanese breweries use this type of yeast," he wrote in *Saké:*

The Essence of 2000 Years of Japanese Wisdom Gained from Brewing Alcoholic Beverages from Rice. "The success of this research was one of the high points of my life."

Today, the domestication of yeast — and, presumably, its domestication of us — continues. We fund the machinery of biology to understand yeast better, because doing so allows scientists to understand us humans, too. We build entire infrastructures, like the yeast rooms in big distilleries or the collections like the NCYC, to preserve and protect the yeasts we love. Yeast, absent any kind of intelligence, inspired us to build a civilization.

Two

SUGAR

IN 1853, COMMODORE Matthew Perry sailed into Tokyo Bay and opened diplomatic and trade relations with Japan under threat of war. Until then, Japan had been almost xenophobically isolationist, but Perry's arrival forced a new approach—the Japanese would have to find ways to relate to a world with which they'd had little contact.

How different was Japan from the West? Well, like most humans, many Japanese people enjoyed an occasional alcoholic beverage, and those beverages were made using yeast. But in Europe and the New World, the substrate, the thing the yeast fermented, was different. The *gaijin*—that's "foreigners," literally "outside-people"—used fruit or grains. In Japan, they used rice.

Yeasts consume sugar—but nature makes a lot of different kinds of sugars, and yeasts don't eat them all. *S. cerevisiae* readily digests simple sugars in most fruits, but grains pose a separate challenge. Their sugars are mostly locked into polymers, molecular architectures with sugar as the basic unit, like a Lego brick. Starch is one; cellulose, as in wood and paper, is another. Yeasts can't break the Legos apart; they can't get at the simpler sugar subunits to make a meal of them.

The existence of beer (made from grain) and sake (made from rice) prove that both cultures — Asia and the West — solved that problem. But they did it with radically different approaches, two roughly parallel developments that probably say more about the centrality of booze-making technology and of sugar itself, as a molecule, than they do about cultural differences. The fact is, no matter where they lived, if human beings wanted to make booze — and they really, really did — they had to figure out how to crack starch.

Half a century after Perry arrived in Japan, a young chemist nearly managed to bring the Asian solution to the Western world. In the process, he discovered a bunch of important things about sugar. And he nearly upended the world of alcohol in the process.

The chemist, Jokichi Takamine, was born in Takaoka a year after Perry arrived and grew up in what's now Kanazawa. His father was a doctor with an interest in the West unusual for the time — he could speak Dutch, for example. Takamine's mother's family owned a sake brewery. The new openness of the Perry era inspired the Lord of Kaga, for whom Takamine's father worked, to send a delegation of youngsters from his province to check out the *gaijin* in the "open" port city of Nagasaki, 600 miles away. Takamine, twelve years old, went to live with European families to learn English. He was one of the first students at the newly chartered University of Tokyo, and then in his twenties, he went on a Japanese government–sponsored trip to Scotland. All this international experience would have made him, back then, one of the most worldly Japanese kids on earth.

Returning to Tokyo in 1883, Takamine got a job with the Department of Agriculture and Commerce. He was supposed to figure out how to industrialize and scale up indigenous Japanese industries for an export market, and he hit upon sake, though his biographers don't know how Takamine got interested. His chemistry professor in college was an inorganic chemist who wouldn't have cared at all about enzymes or brewing. Takamine would have had access to R. W. Atkinson's 1881 book *The Chemistry of Sake Brewing,* one of the first scientific dives into the drink, but that doesn't mean he read it. Oskar Korschelt, author of the 1878 article "Über Sake," was a lecturer at the university, but apparently didn't teach Takamine.

Maybe it was a combination of his mother's family experience and the fact that sake is one of the foods that defines Japanese culture. In the local language it's called *nihonshu,* which means, simply, "the national drink." It's made from rice, equally central. (In Japanese, "rice" is *gohan,* which just means "food.") Most important, though, sake requires another ingredient in addition to yeast, a fungus called koji.

Koji is at the core of Japanese cuisine — it's the key to sake, soy sauce, miso (the fermented soy paste used as a basis for soup), vinegar, and tofu. Technically, it's the fungus *Aspergillus oryzae,* and if you were an infectious disease specialist, that would freak you out a little bit because most of the *Aspergillus* genus is poisonous. In humans, *A. fumigatus* causes the disease aspergillosis, characterized by a serious allergic reaction, pneumonia, and sometimes a bleeding "fungus ball" in the lungs. Several *Aspergillus* species secrete aflatoxins, making the grain they infect — often corn — both toxic and carcinogenic. Some of koji's cousins, in other words, are monsters.

Koji, though, is a pussycat. Just like yeast, it's a tame microorganism at the center of a process with huge cultural and economic signficance. And just like yeast, it was a mystery. It shows up in Chinese records as early as 300 BC, and in Japanese records in 725. Two hundred years after that — which is to say, 1,000 years ago — making and selling koji was a thriving business in Japan, where businesses called *moyashi* have sold *A. oryzae* since the thirteenth century.

Koji does something that sounds simple, but is actually a little miraculous: It turns starch into sugar. Takamine didn't know how — nobody did — but he knew that if he could work it out, it could make him rich.

Sugar is the most important molecule on earth.

You may think water deserves this title. I get it. Water is really good at dissolving other molecules and carrying them around, both inside our bodies and out in the world. Water lets chemicals bump into each other and then do interesting things. But calling water the Best Molecule is like giving paper the award for Best Book. Water is the medium, a backdrop. Sugar is fuel. It's the gasoline in our tanks, the molecule that stores the energy that we living things use to stay alive. (And unlike *actual* gasoline, sugar is water soluble, so our water-filled bodies can move it around easily.)

Fundamentally, sugar is *power*, energy stored in the bonds that hold the molecule together. It's a carbohydrate, which means that it's made of carbon atoms — organized in a pentagon or hexagon, usually — studded with hydrogen and oxygen. From an evolutionary perspective, the reason animal brains think "sweet" equals "delicious" is that we associate molecules that have these structures and ingredients with the idea of a lot of calories for only a little effort. "Sweet" is the brain's reward mechanism for eating energetically dense food.

Sweetness lets plants dangle sugar as bait, loading fruit with it to attract the animals that spread around their pollen and seeds. Honey is loaded with sugar because honey is food for baby bees, and babies need a lot of energy.

The simplest sugar molecules are the monosaccharides. Glucose is a ring of six carbons. Fructose is a ring of five or a ring of six. Stick those molecules together and you get slightly more complex sucrose, or table sugar. Yeasts can eat all those, as well as some of the more exotic sugars — maltose, melibiose, lactose, galactose . . . you get the idea.

Sugar can also be *structure*. Sugar molecules combine and connect. You've heard of the genetic material DNA — that's deoxyribonucleic acid, the backbone of which is the sugar ribose. Form glucose molecules into sheets, sandwich them together, and you get the most common organic molecule on the planet, super-tough cellulose. Glucose is the brick; cellulose is the wall. But where glucose is a basic energy source for most life on earth, cellulose is completely indigestible to all but a few living things. The multiple stomachs of ungulates like cows harbor microorganisms that make the enzymes to break cellulose down. Termites do the microorganism trick, too. Rabbits poop out undigested cellulose just like we humans do, but then rabbits eat the poop to take another digestive pass. Some fungi, too, love the stuff —but not yeast.

Now, here's the cool part: Attach those glucose blocks with a slightly different bond, and you get a whole other material. Instead of tough, indigestible cellulose, it's amylose, more commonly known as starch. Add another twist, and you have amylopectin, another common constituent of plants. It's a super-elegant move by nature, using and reusing the same Lego bricks, again and again—as energy and structure, fuel and walls. That's why sugar is so central.

Which brings us to the problem: We humans can digest starch. Yeasts can't. Without simple sugars, there's no fermentation, and without fermentation there's no alcohol. So this is where yeasts, tricky little bastards that they are, have managed to teach us a few tricks. Remember those dogs acting all juvenile and cute in return for easy meat and a place at the fire? Yeasts learned to roll over and play fetch, too—in return for access to sugars they couldn't get at on their own. We learned to break apart the complex sugar polymers in grain so we could feed them to a fungus. We domesticated yeasts; yeasts domesticated us.

It didn't have to go this way, of course. Almost everywhere in the world, people use the simple sugars to make alcohol. You can get it from sugar beets like Monsieur Bigo in Lille, and from a New World grass called sugar cane. Ferment and distill molasses, a byproduct of turning cane into table sugar, and it's rum; use straight cane juice and you get a funkier, weirder drink called rhum agricole.

You could also use honey, if you have it, to make mead. If you have neither cane nor beets nor honey? Well, since at least the thirteenth century the people of the Central Asian steppes have made *koumiss* from horse milk (it has more fermentable lactose than milk from cows or goats). In Sudan, they use milk from camels. Lots of cultures use tree saps as substrates; maple is an obvious one in the West, and in Africa they use date-palm fruit and sap. Those substrates have a needle-burying 60 to 70 percent sugar content and plenty of ride-along yeast species. Palm wine is *asante* in Ghana, or *nsafufuo* or *ewe.* Nigerians call it *ogogoro;* South Africans, *ubusulu.* Expose straight palm sap to air, and wild yeasts—as well as lactic acid bacteria and whatever else is floating around—begin fermentation right away. In less than a day tappers can sell the resulting brew in gourds or recycled jars, either to local bars or by the side of the road. The wild yeasts floc on top in snotty strings and clots; lactic and acetic acid producers in the local microflora produce flavors like spoiled milk and vinegar, evolving into a potable that, by one account, tastes like eggs and crocodile fat.

Agave—a succulent, not a cactus—grows in deserts across the Americas. A single plant can hold 50 to 250 gallons of glucose, fructose, and sucrose-filled sap. Ferment that and you get pulque, a sweet-tart drink that's also famously snotty—bacteria associated with the yeast form a viscous sludge called a biofilm during manufacture. The main carbohydrate in agave is a fiber called inulin; if you cut the leaves off and roast what's left, called the *piña,* the inulin breaks into fructose molecules, perfectly good food for yeasts. Ferment and distill goop made from that and you get tequila.

(Sometimes. Technically you have to use *Agave tequilana* Weber var. *Azul*—that's blue agave—and you have to make it in the Tequila region of Mexico. "Tequila," it turns out, is a controlled appellation, like "Cognac" or "Bourbon." The regulations say you have to follow certain recipe specifications *and* be in the right place. And if you try to make the same thing somewhere else? You're making a different drink. Use *Agave potatorum* and you're making mescal. Mix in some fruit and a roasted chicken, and you're distilling *pechuga,* which fortunately does not taste like chicken.)

Or why not, you are screaming by now, just use fruit? And indeed, almost every culture does. In early America, colonists fermented just about everything they could find—pumpkins, maple, persimmons, and especially apples. Cider was the premier social lubricant until Dutch and German immigrants—with their brewing heritage—discovered that barley would thrive in Pennsylvania.

So let's build the ideal booze substrate. We need a fruit that grows widely and easily, with a high sugar content. It should maybe have some interesting flavors, or be easily manipulated into having interesting flavors. It should be easy to harvest and easy to ferment.

What we're looking for is the grape.

One way to think about sugar is that it's the way living things store and move carbon. When biologists and science fiction nerds talk about "carbon-based life forms," this is what they mean. So tomatoes, for example, store carbon mostly as sucrose. Apples use a sugar-alcohol combination. Avocados store carbon not as a sugar but as fat — just like animals do. But grapes are right near the top among fruits in terms of storing carbon as simple, monosaccharide sugars. The fruit is one-quarter sugar, and half of that is glucose.

But grapes are about more than sugar. A lot of what we think of as flavor comes from volatile chemicals, molecules light enough to vaporize and turn into something our noses can pick up. Most fruits — like, say, apples — make a lot of volatile compounds, especially in the form of the alcohol-acid combinations called esters. But grapes "don't make a lot of esters at all," says Paul Boss, a plant molecular biologist with the Commonwealth Scientific and Industrial Research Organisation in Adelaide, Australia. That's good news for winemaking, because the chemical process of fermentation would destroy whatever esters the grape came up with. But the grape makes molecules with the potential to *become* esters when the juice gets turned into wine. "I guess when ancient man was experimenting with making alcoholic beverages from whatever he could lay his hands on, that's why he didn't pick tomatoes," Boss says.

Human settlers of the temperate, well-watered, fertile land around the Tigris and Euphrates Rivers had access to a lot of fruit — olives, figs, dates. But of all those old-school sugar sources, only the grape stores most of its sugars in simpler, soluble forms — as something that yeasts could readily metabolize. In fact, once you can settle down long enough to collect a bunch of grapes, it's actually hard *not* to make wine. Just bruise the grapes and come back later. They'll ferment on the vine. Domesticated grapes, the grapes we use for wine today, make all the right molecules, grow to the right size, achieve the right Brix (the unit winemakers use to measure sugar), and in general sit and roll over on command.

Wherever there's a culture of wine drinking, one single species of grape dominates: *Vitis vinifera.* In his book *The Chemistry and Biology of Winemaking,* Ian Hornsey suggests that the main reason is topography. Mountain ranges in the Americas and in eastern Asia run mostly north–south, Hornsey says, but in Europe and western Asia they run east–west. So as Ice Age glaciers spread southward, grape species in the Americas and China could go on the run, as it were, fleeing south for warmer, friendlier climates. But in Eurasia, the grapes had to find tiny "refugia," safe microclimates where they could hide out and wait for the thaw, about 8,000 years ago. Only *V. vinifera* made it out alive.

North and Central America have thirty species of *Vitis;* China has another thirty. But Eurasia has only *V. vinifera.* And Eurasia is where wine was born. From Chardonnay to Pinot Noir, Syrah to Viognier, all wine — no matter what the label says, what color it is, or where it comes from — is made from the same species of grape, probably originating somewhere in the Transcaucasian Highland connecting the Black and Caspian Seas, in what's now the country Georgia, and then spreading south to the Fertile Crescent and Egypt. (Recent genetic research suggests at least two domestication events for grapes, much like the multiple moments when human beings turned wild yeast into something useful. One might have been in Transcaucasia, but the other was probably in the western Mediterranean, giving rise to western European grape varietals — what botanists call "cultivars.")

So what gave the grapes such an aptitude for booze? They were easy, for one thing. Grapes grow in places other plants can't, in soil other plants can't use. Their tendrils let them climb on other crops, or amid scrub or trees. They grow extended vines called lianes, but they can survive pruning.

Their chemistry, too, is perfect. Most of the fruit is pulp, or what botanists call mesocarp, and most of that pulp is made of the sugars glucose and fructose, as well as tartaric and malic acid, a yeast-friendly mix. Much of the grape's flavor comes from volatile aromatic compounds called terpenes — like geraniol, which smells like geraniums, and linalool, which is kind of spicy-floral. Many plants make and emit terpenes, but most of them do it with specialized structures, like the little projections on peppermint leaves, called trichomes, that give off menthol, or the glandular pockets that sequester oils in citrus fruit peel. But in grapes, all those yummy-smelling molecules kind of float around in the pulp — and eventually in wine.

All grapes have clear juice, but the skins are full of pigments called anthocyanins. (Macerating the skins in the juice makes wine red.) Those pigments are also full of big, astringent polymers called tannins and compounds with a chemical attachment called a phenol group — that's going to be important later, but it's basically the smell of coal tar or oil, and it's one of the things winemakers measure to determine ripeness.

So wild grapes are interesting packages of flavors that helpfully signal when they're ripe. As grapes grow, they're small, greenish, and acidic. Then they go through a phase vintners call "veraison," when they soften and turn reddish. In a final phase, they get bigger and accumulate sugars, an evolutionary trick that ensures that the animals that eat them and distribute their seeds don't do it too soon. The seeds aren't ready to be spread until the fruit is at its sweetest and most attractive to birds and other connoisseurs.

So, sounds perfect, right? Domestication made grapes even better. Human beings forced *V. vinifera* into a liberal redefinition of sex roles. Wild grapes are either male or female, with insects or birds or bats or the wind carrying pollen from the male plants' flowers to the flowers of the female plants, which bear the grapes. But that makes propagating the traits you like — bigger berries, a recessive tendency to be greenish-white instead of blackish-red, and so on — difficult. You have to keep the female plants separate from the male plants, and you risk losing the target trait with every new cross.

The solution? Make the plants intersex. Wild grapes have both male and female plants; just like animals, some plants reproduce via the exchange of genetic material between the two sexes. That may be the least fun way to think about that process, but the point of it is to swap genes around, to ensure diversity within the species. That's great if you're trying to evolve and adapt, but it's terrible for the person trying to grow and harvest you because of whatever skill set you've settled on. That person doesn't want you to change.

So one of the hallmarks of grape domestication is a switch from being dioecious — having a male and a female — to hermaphroditism. As Sean Myles, a geneticist at Dalhousie University in Nova Scotia, explains it, hermaphroditic grapes can fertilize themselves; flowers are more likely to get pollinated and turn into grape berries, resulting in full clusters. In the wild, dioecious ancestor of the domesticated grape, only female plants produce fruit. That's where all the different grape varietals come from, each one customized to the needs of the winemakers. They're all *V. vinifera,* but look at the variety. Just one region in France, the Haut-Médoc, grows the low-sugar, high-tannin Cabernet Sauvignon (good for aging) and Merlot, with the exact opposite traits (good for elevated alcohol levels). Vineyards there also grow early-ripening Cabernet Franc and late-ripening Petit Verdot, with its higher levels of tannins and sugar.

For someone with a contrarian streak, the existence of all those cultivars suggests that grapes weren't the perfect ingredient for booze-making we've been led to believe. If grapes are so great, why did we need so many different custom versions? That's the less Panglossian take on grape domestication Myles offers. "They happened to be the juiciest fruit in the Fertile Crescent when everything was getting domesticated," he says. "If the world's primary center of cultural development had been in Melanesia it could be that we'd all be sipping on fermented coconut juice and we'd have all these different coconut cultivars with different profiles." Grapes: What's the big deal?

"These cultivars are frozen in space and time, and the pathogens around them continue to evolve, so the growers bow down to the chemical companies," Myles says. "We shouldn't be planting old ones. We should be planting new ones. We should be shuffling up new genetic combinations that are superior."

He studied the grape genome as a postdoctoral researcher at Cornell University and advocated for the creation of more advanced, newer strains through breeding — just as you'd find with any other commercial fruit or vegetable. But his work in grape genetics went for naught. It takes a long time to develop a new strain, and Myles couldn't sell anyone on disease-resistant, more flavorful grapes. The winemakers of Europe and California only wanted the same dozen or so strains, familiar from labels. It was "grape racism," Myles says. So he got out. "I'm in the apple game now," he says. "We can release a new cultivar, and if it does well, we can make a killing."

In 1884, Jokichi Takamine came to the United States, a delegate to the World's Fair in New Orleans. Back then, World's Fairs were more than just pop-up theme parks; they were combinations of natural history museums, art galleries, and global trade expositions that lasted for months at a time. It was a good gig.

Takamine rented an apartment in the French Quarter and met Caroline Hitch, the beautiful blond, eighteen-year-old daughter of his landlord, a retired colonel in the Union Army. Though Takamine was almost twice Caroline's age, the two got engaged. It might have been a bit of a scandal, but as exotic as Takamine must have seemed, his English was good and he didn't rush things. At the fair, he learned about fertilizer made from phosphate, a relatively new industrial process that made for tremendous improvements in crop yield. When the fair was over, he went to Charleston, South Carolina, to learn how it worked, and then went home to work for the Department of Agriculture. He ran the country's patent bureau for a year, and eventually took charge of the newly created Tokyo Artificial Fertilizer Company. In the summer of 1887, Takamine returned to New Orleans to marry Caroline. They moved to Tokyo.

Takamine kept working on koji. He was trying to figure out a way to improve the process, to speed it up and to extract "diastase," whatever the stuff was that actually converted the starch to sugar. Eventually Takamine found a way to grow koji mold not on rice but on wheat bran, up until then a useless waste product. He called the new product, perhaps self-aggrandizingly, "taka-koji." Takamine also figured out how to use alcohol to extract the active ingredient, and then how to powder it. He had isolated starch-breaking enzymes, even though no one yet knew what an enzyme was.

Takamine realized he had more than a new way to work with koji. He'd discovered a new way to make ethanol—and his approach only took forty-eight hours, versus the several days it takes to saccharify barley. The established process, called malting—as in single malt whisky, malt liquor, malted barley, and malted milk shakes—had been the Western approach for thousands of years. And Takamine could beat it. Pound for pound, Takamine's extract actually produced more sugar than malting—which is like more money to brewers and distillers.

This is where things could have changed. Takamine had what he was looking for—a cheaper, faster process to turn starch from grain into fermentable sugar. All that remained was to try it on an industrial scale. In 1890, Takamine got his chance. His father-in-law sent him a telegram telling him that the biggest distilling conglomerate in the United States wanted him to come to Chicago and try to make whisky without malt. It would have been a massive uprooting of his growing family, back to the United States—and to the Midwest—from Japan. The fertilizer company was doing well . . . but Caroline was not. "Her mother-in-law could not stand her," says Joan Bennett, a Rutgers University *Aspergillus* researcher and Takamine biographer. "She really was quite miserable. That's probably why they returned to the US."

Scientists at the time typically didn't think much about the commercial implications of their work, but Takamine had no such limitations. Takamine, Caroline, and their two children moved to Chicago, and Takamine built a demonstration facility. In 1891, he took out a US patent on his process—depending on how you count such things, it was the first English-language biotechnology patent. And that same year the Whiskey Trust, a major boozemaking conglomerate, hired Takamine and moved him to Peoria so he could build a full-scale industrial facility. As the *Chicago Daily Tribune* reported from Peoria, "The Distilling and Cattle Feeding company of this city, which controls the whisky manufacture of the world, has adopted a new process in the manufacture of spirits which will yield better returns than a gold mine." Takamine was set to get a fifth of whatever profits the distillers realized.

He was on the cusp of a new world of booze. But the Old World wasn't quite ready to let go.

The small village of Muir of Ord, far in Scotland's north, has a rail station and surprisingly wide roads — built for trucks. That's because the town is the gateway to the Black Isle, which despite its foreboding name is one of Scotland's prime barley-growing regions.

Muir of Ord has had a whisky distillery since 1838. Today it's a boring, squared-off building owned by the transnational drinks conglomerate Diageo. A smaller, wooden building with peaked roofs next door houses a little shop which I have read is popular and I assume is charming, but I can't say for sure, because by the time I arrive at Glen Ord one overcast Friday evening, the distillery is closed up tight, doors locked, lights off. And that's fine. I've never tasted the whisky they make, the Singleton of Glen Ord; most of it gets blended into the megabrand Johnnie Walker, and the rest is sold almost entirely in Malaysia and Thailand. Here's the review from one whisky blogger: "A whisky that is pleasant enough — but nothing special nor memorable for being either great or crap."

What's important about Glen Ord is the other building, across the parking lot. It's a sixty-foot-high Modernist cube attached to another cube. This is Glen Ord Maltings, one of four facilities around Scotland where Diageo turns barley from a starchy grain into sugary malt, ready to be turned to whisky. Eleven people work here, every day, processing 38,000 tonnes a year of barley. That's 83.8 million pounds, seven or eight trucks a day arriving from all over Scotland and then shipping back out to Diageo distilleries all over the country.

Distilleries used to malt their own barley in-house. It was what defined a whisky, in part, as a "single malt." The characteristic ziggurat-shaped roof that today identifies a whisky distillery is actually the roof of the "malting floor," where the maltmen would bring in barley and get it damp, turning it over with specially made tools that look like hand plows to control the heat generated by germination, and then arresting the process with a gentle cooking step (sometimes using peat moss as fuel) to turn it into malt. But few still do it. The malting floor was beautiful, but it's a relic, more expensive than using a central facility like Glen Ord, and today, with global demand for single malt skyrocketing, even a football field–sized malting floor wouldn't allow a large distillery to cover its own capacity. A malting floor takes five or six days to malt 10 to 12 tonnes of barley, and a normal-sized whisky distillery eats that much barley in a single batch.

Malting is how people turn barley's starch into sugar, which they can turn into beer and booze. Whisky is, basically, distilled beer — and this was the process Takamine hoped to obliterate.

Brewers and whisky makers use barley because it's easy. Other grains are maltable, but wheat, for example, produces less starch-breaking enzyme. Oats have too much protein and fat. Corn needs too much heat to untangle the starches before malting, and the oils tend to turn rancid. Barley is the way to go.

Even though it's Friday after hours, Daniel Cant, Glen Ord's site operations manager, is still at work. He has dark, short-cropped hair and an unlined face, and he's wearing a reflective orange jacket. He's carrying a vest of the same color for me — we are, after all, about to visit a working factory floor, a place where 100-million-year-old biochemistry has been harnessed as an industrial process.

At the loading dock, a dump truck has its trailer tipped up, and from a pipe in the back a stream of barley as wide as a weightlifter's thigh pours into a hole in the concrete. Cant dips a hand into the flow and carries a small handful of seeds into his mouth, inviting me to do the same. I take a bite. It's like dry breakfast cereal, not particularly sweet. Think Grape-Nuts, not Lucky Charms.

"Now, look at this," Cant says. He puts a single grain between his thumbnail and index finger and presses down, hard. His nail barely makes a dent; the grain's husk is tough. After it drops into an underground hopper, it'll get carried via conveyor belt past a magnet (to pull out any metal, like farm machine parts or Roman coins or whatever else harvesters accidentally pick up on millennia-old Scottish cropland). Then the barley goes to one of twenty-five silos on-site, each capable of holding 200 tonnes.

Inside the plant, the grain rumbles and shushes through crisscrossing vents and ducts all around us, resonating through the walls and the metal-mesh floors of the factory. Cant and I have to half shout at each other to be heard. "Distilleries, it's easy. You're moving liquid around," Cant says. "Grain is harder." A facility like this poses a massive problem in materials handling. Particles in motion sometimes behave like liquids but sometimes behave like sand, creating little tumbling avalanches instead of steady flows. Instead of valves and pumps, the ducts use turbines, paddles, and conveyor belts, which the grain constantly wants to clog. If dust disperses through the air it behaves like an explosive gas — any spark can ignite a particle, which then sets fire to all the particles near it, and so on, in a three-dimensional, fast-moving exothermic wave, which is a fancy way of saying "fiery death explosion." So a malting facility like Glen Ord is studded with vacuum machinery, like a good carpentry shop. (When the grain gets wet, all those problems get replaced by the challenges of moving sticky, heavy sludge.)

To get to where the actual malting happens, Cant slides open a giant metal red door to reveal what looks like the boiler room for a battleship, circa 1964. We climb stairs to the highest floor, where hoppers drop dry barley into eighteen wide pools of water called steeping vats, each one big enough to swallow a Mini Cooper. Through the floor, an open metal grate, I can see twenty feet down to the tops of eighteen cylindrical metal tanks as big as tanker-truck trailers. Their faces are studded with gear teeth, like they should be operated by hapless underground workers in *Metropolis.* They are all filled with wet barley, turning slowly.

Where metal cross-braces intersect above the vats, tufts of turf grow where barley grains have gone astray and settled. "This is where we get the moisture into the barley," Cant says next to a vat. The grain goes in at about 13 percent moisture, so dry that it can last for a couple of years in a silo. Two days after the barley first drops into the tank, it has up to 48 percent water content. That's the trigger for a chemical cascade.

Barley, like all grains, is a grass, and the part that we humans eat is its seed. A seed is a bomb of life, an embryo packaged with all the nutrition it will need to be born and a biochemical engine for growth and development. Drop that bomb into water and air and dirt, and it detonates into a plant. But if we're interested in alcohol, we're interested not in the payload—the rootlets and embryo that turn into a plant—but in the packaging. Those seeds are full of starch.

A barleycorn is as streamlined as a tuna. It grows with its "head" pointing inward, toward the stalk, and at the tip where the tuna's brain would be is the barley embryo, a packed bundle of rootlets. Behind that is a wall called a scutellum, which is what turns into leaves. And behind that, taking up almost the entire space of the corn, is the starchy endosperm, food for the embryo (just as the yolk of an egg provides nutrition to a developing bird). Wrapped around that are three layers of enzyme-making cells, the aleuron, and the whole thing is encased in a hard cellulose husk. That's why Cant's thumb could barely dent it. Cellulose is tough.

All that handling of barley, getting it soaking-wet and then air-drying it, tricks the barley embryo into thinking that it's time to start growing. It triggers a chemical cascade, starting with the production of a hormone called gibberellic acid, which then courses through the aleuron layer. That's a signal to those cells that they should start making enzymes—amylases that can break down starches and proteases that can break down the protein coatings around those starches. Those amylases can break down not just the starch in the barley but in anything—maize (the main component of American-style bourbon whiskies), sweet potato, and even the rice used as an adjunct grain in some beer.

Cant and I climb down a metal staircase to one of the turning drums. He taps a few keys on the control panel to make sure it's locked and won't start turning while we have our heads inside it, and then steps down to a square metal door in its side and opens it up. The smell of fresh, vegetal cereal puffs out, like Midwestern farmland in autumn. That's a good sign — if the grain was too wet, that scent would be more like apples, like fermentation.

He scoops a few grains out; now his thumbnail cuts right through an individual grain. Inside is a curving stripe of luminous white called the acrospires. It's the beginning of the rootlet, where the modification of starch into sugar begins. "You want the enzymes to break down the starch and protein, but you don't want them to start growing the plant," Cant says. In other words, you want the seed to make sugar but not eat it. Cant pushes on a seed and it crumbles into white bits. "It should feel like icing sugar," he says. It smears across his fingertips, a thin white streak.

From there, Cant has to dry the barley again. Wet, it can get infected by molds and fungi, and aside from being potentially toxic, their flavors make it all the way through to the end product of distilling. Also, all those different colors and roast levels that brewers like — "chocolate malts," and so on — get introduced through heating. So Cant kiln-dries it, using waste heat from the distillery across the parking lot or an oil furnace. "You can also add phenol," Cant says. "As in peat."

That's a big deal for whisky makers. A mélange of sphagnum moss and other plants, peat forms when the tangled combination dies underwater, in marshland or bogs. Absent oxygen, the bacteria that ordinarily break down dead plant matter can't get at it. So it accumulates, and in Britain it was, for a while, a major fuel source — cut out of the bogs, dried into bricks, and burned. But the resulting smoke had a particular flavor, from the phenolic compounds in the moss. People who like it tend to call it names like "earthy" or "iodine-like"; people who don't like it say it tastes like old Band-Aids. Glen Ord runs through 38 tonnes of peat pellets every week; there's a huge pile of the stuff outside. Every distiller wants their parts-per-million phenol count a little different. At Glen Ord they make a batch at 100 parts per million — that's high — and then mix it with unpeated malt to lower it back down to a given distillery's spec.

Back in the office, Cant shows me a dorm refrigerator full of what look like plastic sour-cream tubs. They sport labels that say things like "Delivered Malt — Cragganmore" and are full of samples of what they've sent to the various distilleries Glen Ord serves. Cant pulls one out, tells me it's a test batch of a newer strain of barley called Concerto, and offers me a taste — my third sample today.

It's crunchy, like a Japanese rice cracker, and the sweetness has mellowed from something Cheerios-like to porridge with syrup. The phenols from the peat ring out clearly; it's medicinal in that way that seems lovely if you're the kind of person who thinks that peat is lovely.

It's finished, in other words. It's waiting to be whisky.

Sake makers work with about forty different strains of rice — *Oryza sativa,* for you taxonomy nerds — and like barley strains, each is optimized for various traits. For example, Hiroichi Akiyama, the sake maker I mentioned in the last chapter, describes the strain Yamadanishiki as being more than a little finicky. It only grows in the mountains, which makes it hard to cultivate and harvest — big agricultural machines can't handle the slopes. It grows tall and ripens late, which means there's always a risk that it'll get blown over by typhoons before the harvest. But it's worth it because Yamadanishiki, like all the great brewing-rice strains, grows large grains with a lot of starch on board and little protein.

Rice has a tough, armored husk, and just inside it a layer of germ — those, together, are bran, which makes brown rice brown. For sake brewers, bran is trouble — the proteins and fats give sake all sorts of off flavors and colors, and they inhibit the growth of yeast. So they mill away that bran layer, and more. The process is called "polishing," and it involves dropping uncooked rice onto a roller coated with super-hard silicon carbide that grinds away the bran, the enzyme-making aleuron layer, and a bit of the starch at the grain's heart. That polished rice gets steamed, filling a sake brewery with one of the most comforting smells in the world, a heavy, sweet, nutty aroma. Then the rice cools a bit . . . and the real problem emerges. Steaming softens the starch — "gelatinizing" it — but doesn't break it down.

In the early days of sake, people broke the starch by chewing the rice and spitting it out. Human saliva contains amylase; spit starts breaking down food well before it reaches our stomachs. In fact, in a remarkable example of parallel cultural evolution, people in rural South America make a fermented beverage called *chicha* the same way, chewing manioc or cornmeal and then taking it out of their mouths, rolling it into little lumps, and sun-drying them before fermentation.

Expectoration is not a long-term, scalable solution to this problem. For rice wines, the solution ended up being even weirder. It was koji.

Just like yeast, nobody isolated and identified koji until the late 1800s —1876, to be specific. Yeast was the first living organism to have its genes sequenced, in 1996; koji didn't get its turn until 2005. What gene jockeys found was a microorganism 20 million years old, yet suited to a thoroughly modern process. It makes ten proteases, beloved by soy sauce and miso makers for breaking down protein-laden soybeans; and it makes three distinct alpha-amylases, which sake brewers depend on to saccharify rice.

Koji still has the genes that make its cousins dangerous. Genetically, it's 99.5 percent the same as the aflatoxin producer *A. flavus.* Yet *A. oryzae* seems perfectly safe. "Those genes that might have bad effects were almost completely repressed," says Masayuki Machida, head of the Molecular Systems Bioengineering Research Group at Japan's National Institute of Advanced Industrial Science and Technology. He was lead researcher on the *A. oryzae* sequencing project. "In the case of strains used in the fermentation industry," he says, "some genes which might have bad effects were completely deleted." But Machida's work didn't explain how — or when — people tamed the fungus.

Solving the mystery was a longtime goal of the evolutionary biologist Antonis Rokas. When he was putting together his lab at Vanderbilt University in 2007, he asked Japan's National Research Institute for Brewing to send him mold spore samples isolated from various breweries' koji. Rokas wanted to study evolution, and his angle was domestication — really just a human-piloted, accelerated version. "I feel like a lot of our concepts of evolution are heavily biased by examples in research from animals and plants," Rokas says. "That doesn't make it wrong, but a lot of evolution is microbial."

Rokas started thinking about domestication in a new way. In breeding for specific traits, domestication selects for sets of genes. Different versions of genes, called "alleles," make different proteins, and those have different effects. That could mean dark hair, or bigger fruit, or (in the case of domesticated microbes) less bitter beer, or faster-rising bread, or greater production of penicillin.

But no gene is an island; they're chained together and packed into structures called chromosomes. Rokas hypothesized that domestication — that is, selection for a certain trait encoded by a certain gene — also reduced variation among neighboring genes. You get the genes that break down starch in rice, sure, but you also get whatever is sitting next to them on the chromosome. "If you look a few thousand years later, these regions where you have selection, with traits that would have been advantageous, would have reduced variation," says Rokas.

In other words, a domesticated organism would have chromosomes with regions of less variation relative to its wild ancestors. Pieces would be frozen in time. So Rokas's team did an experiment: They spread out the genome of *A. oryzae* and lined it up next to the genome of its toxic cousin *A. flavus,* and went hunting for those frozen regions.

They found them. About 150 stretches of DNA had normal variation in *A. flavus* and a spooky stability in *A. oryzae.* Even better, the genes in those regions were plausible domestication candidates. "A lot of these genes have to do with metabolism, which intuitively makes sense," says Rokas. "If you're in the business of breaking down starch in rice, metabolism would be a big deal." One of the most stable was a gene that codes for a glutaminase, an enzyme that turns the amino acid L-glutamine into glutamic acid. You know this stuff better as part of monosodium glutamate, MSG, the "flavor enhancer" that conveys the meaty, proteinaceous flavor umami. It's a major component of soy sauce, miso, and, yes, sake.

According to Philip Harper, the only non-Japanese *toji,* or master sake brewer, in Japan, koji is the most important thing to understand about sake. It has a cultural impact way beyond the usual microbe. Large-scale breweries infect their rice with koji in "culture machines," massive tanks — the one at Takara Sake in California, a few blocks from my house, is a steel hexagon two stories high, capable of transforming 6,000 kilograms of rice at a time. But Harper's brewery infects the rice in the traditional way, spreading it out in a layer and sprinkling koji spores from a tin cup with a mesh cover; they fall as tiny, yellow-brown particles, half a dozen different strains he buys from three different producers. The smell of the culture room changes as the koji fungus spreads its tendrils through a batch of rice, shifting from a homey, comforting cooked-rice smell to something more like roasted chestnuts or, if it goes a bit longer, the earthy smell of mushrooms. The rice, now also called koji, tastes different — sweeter, as you'd expect, and a little like popcorn. It goes from being translucent and almost opalescent to a solid, luminous white. "When you're making wine, you have a direct line from the grape," Harper says. "The whole point of sake making is, you start off with steamed rice, which is one thing, and with two days in the culture room you have not rice but rice koji, which is a completely different entity. You've had a radical transformation even before you start thinking about fermentation."

If you could apply the koji-transformation process to barley, you wouldn't need malting at all. And you could really speed things up if you didn't even need the koji — if you could just use the enzymes it made. Today you can buy those, but in the late 1800s, you couldn't. That's what Takamine was offering: starch breakdown without malting.

You know who didn't like this idea at all? The maltmen. Even as Takamine tried to industrialize his process, the malt manufacturers campaigned against the project. In early October 1891, Takamine suffered his first real setback: In the middle of the night, his employer's distillery caught fire. It was . . . suspicious. Here's how the *Peoria Daily Transcript* played it:

> The fire at the Manhattan malt house early yesterday morning was a most peculiar one and it is only by sheer luck that other buildings were not destroyed. When Hose Co. No. 6, at whose house the alarm was turned in, reached the scene the fire, though blazing brilliantly, was confined to a small frame tower, and under favorable circum-

> stances could easily have been extinguished. They laid a line of hose, but so great was the distance from the burning building to the nearest hydrant that the hose would not reach, and the crew had to stand about until the arrival of Hose No. 4, who completed their line. The water was then turned on, but to their extreme disgust the firemen found there was no pressure.

Hmm. That's funny.

Takamine tried to rebuild, and finally got a distillery functioning with barley saccharified with taka-koji. It took him three years, but eventually he was converting 3,000 bushels of corn a day. His booze went to market—a cheap, malt-free whisky called Bonzai. But it never really took off, and eventually his relationship with the Whiskey Trust soured. The business degenerated into a long legal battle, and the Takamines' money ran out. Caroline had to sell her art collection. They had to ask their families for money to survive.

In 1894, the Whiskey Trust broke it off with Takamine. He'd continue to advocate for the replacement of malt with taka-koji for decades, but he'd never get a business off the ground to make it work. Malting remains the main way to get the enzymes that convert starch to fermentable sugar.

Think what the boozemaking world would have been like if Takamine had succeeded: The malting facility in Muir of Ord—the entire infrastructure of malted grain, potentially—would be unnecessary. Different strains of barley, different grains altogether, might dominate the brown liquor market. The Asian markets for whiskies that are so lucrative today might have cropped up 150 years sooner, and research into enzymes might have gone in an entirely different—and more commercial—direction.

Don't feel too bad for Takamine, though. He failed at making booze, but he didn't *fail.* Takamine switched his attention to the pharmaceutical business, dubbing his diastatic extract Taka-Diastase and marketing it for "dyspepsia." As Bennett puts it, "he basically created the Alka-Seltzer of the 1890s." It was so successful that Detroit-based pharma company Parke-Davis took over its manufacture and marketing and set Takamine up in a lab in New York to look into another wonder chemical that no one had quite been able to isolate: epinephrine. It seemed to be able to restore the heartbeats of the nearly dead, confer energy boosts, and help with allergies. But no one knew how to make it.

So Takamine went to visit the Johns Hopkins lab of John Jacob Abel, who was also working on the problem, and hired a chemist named Keizo Uenaka to come work with him in New York. Takamine combined Abel's methods with what he knew about purification, and one night in 1900, Uenaka produced a pure crystal of extract.

Takamine filed a patent on the new substance, which he named "adrenaline," and in 1901 published two single-author papers about it. It was all kind of prickish and dicey where Uenaka's contribution was concerned, and eventually a rival pharmaceutical company challenged Parke-Davis's rights to exclusivity in court, saying that Abel had gotten there first, and anyway no one should be allowed to patent a naturally occurring substance. In 1911 Judge Learned Hand ruled otherwise, saying that such patents were legal. His opinion paved the way for the modern pharmaceutical and biotechnology industries.

Between Taka-Diastase and adrenaline, Takamine got rich again. He and Caroline built a five-story mansion in Manhattan and decorated the first two floors in authentic Japanese styles. He started a couple of companies in Japan and the United States, helped found the Japanese equivalent of the National Science Foundation, and in 1912 paid for the 3,020 cherry trees that still decorate the Tidal Basin in Washington, D.C. In more modern times, there have been children's books and biographical movies about him in Japan.

Weirdly, Takamine's vision of a maltless distilling future eventually came true without him, sort of. Corn syrup, that bête noire of some of today's anti-obesity activists, is mostly made with enzymes derived from *A. oryzae*. And while it's true that craft beers, the so-called premium level, use twice the malt of mainstream national brands, a few new beers suggest a diverging trend line using not more malt but less. Much less.

Brewers in Japan, for example, make a drink that is almost but not entirely unlike beer, called *happoshu.* The brewers were being taxed based on the amount of malt in their beer, so they figured they could lower their tax burden if instead of using malt they made a wort, the sugary stuff that gets fermented into beer, with synthetic enzymes and a barley extract. They'd evaporate off the water and mix that extract with the normal brew. Technically the amount of malt was lower.

The cheap, almost-beer drink turned out to be really popular. Every major Japanese brewer made a version. Some even figured that if they didn't need malt, they didn't even need barley—they just needed a source of sugar. Peas, for example.

Meanwhile, in Denmark, a brewery called Harboes for a time sold a thirst-quenching lager called Clim8, touting an 8 percent reduction in carbon footprint in its manufacture versus similar beers. The trick? No malt. Harboes made Clim8 with unmalted barley and enzymes purchased from the biotech company Novozymes.

The yeasts don't care where the sugar comes from, as long as it's broken into simple, Lego-brick monosaccharides. They're happy to have helpful human beings bring them a meal, whether through malting, koji, or the judicious application of enzymes. Actually, let's not say they're happy. Let's say *sated.* And yeasts show their satisfaction by fermenting—by making booze.

Three

FERMENTATION

AT THE END of a shelf in the sunny office of Patrick McGovern, scientific director of the University of Pennsylvania Museum's Biomolecular Archaeology Project, is an object arguably more amazing than anything in the big, slick museum downstairs. It's a fragment of ceramic about the size of a pack of cigarettes, and if McGovern is right, this 10,000-year-old square of pottery, a broken, khaki-colored piece of the base of a jar, grooved on the interior and slightly concave, holds evidence of the oldest known human-made fermented beverage on earth.

McGovern has a Santa Claus–like beard and a slow, considered way of talking. He's not the digging-in-dirt kind of archaeologist; mostly he uses laboratory tools to find faint traces of chemicals in ancient artifacts. Alcohol is rarely one of them — it's an evanescent molecule, so light that it evaporates on the time scale of a human lifespan. But the other stuff that goes into a primitive beer or wine persists. That's what he's good at spotting.

It's not easy. You have to be smarter, let's say, than the German researchers who found liquid in a centuries-old container and simply drank the stuff, figuring they'd be able to taste anything interesting. They didn't; all they could taste was water. Anything perceptible in what they'd found had long since vanished. Luckily science has tools to reveal things that human beings can't perceive.

To a chemist, "clay" is a hydrated alumino-silicate — aluminum and sand and water. That's a good mix for grabbing and holding polar molecules, those with a positive or negative charge, including the acids, esters, and other molecules unique to fermented beverages. The trick is figuring out which ones might survive for thousands of years, and what kind of forensics — or, really, biomolecular archaeology — will reveal them. Fermented drinks contain a complicated mix of molecules, and some of them ought to persist over deep time, especially when retained in the matrix of clay.

McGovern didn't begin his career looking for booze. He started out as an expert in royal purple, the mollusk-derived dye that was one of the ancient world's most expensive and sought-after status symbols. A researcher with whom he'd studied Akkadian at Penn brought him in to look at some suspiciously wine-like red residue on the inside of a 5,000-year-old pot from a site called Godin Tepe, in western Iran. A suite of chemical traces — tartaric acid (found in large amounts in some Eurasian grapes) and pine resin (a common wine preservative in the ancient world) —led McGovern to conclude that it was, indeed, wine, an easy 3,000 years older than the oldest known sample at the time, from Roman amphorae aboard ships wrecked off the French Riviera.

Soon McGovern became the go-to researcher when it came to ancient fermented beverages. So when researchers working at Jiahu, a Neolithic village in northern China, found similarly residue-coated jars and perforated basins, they called him. Jiahu was already famous for old and possibly domesticated rice, the oldest playable musical instruments, and what some people think is the oldest example of Chinese pictograms. A fermented beverage wasn't out of the question. McGovern never actually got to the site itself. He just saw pottery sherds. "If you go to the museum in a nearby town," he says, "they're in a storeroom."

Fragments from one of McGovern's research sites were too fragile to make the trip back to his lab. "I did the extraction in a high school chemistry lab," he says. "I had to buy chemicals locally, and they weren't of very high quality." But eventually, McGovern was able to take some samples back to Penn.

With the right equipment, his tests got more serious. McGovern found traces of the chemicals you'd expect in ancient samples of rice and rice wine — a good get, but not conclusive, since Jiahu is full of ancient rice. Organic molecules called *n*-alkanes, specific to beeswax, were also present. That's a sure sign of honey — sugars degrade, but beeswax is tough to filter out. McGovern also found signs of tree resins, and he found evidence of tartaric acid. That might have been from Eurasian grapes — *Vitis vinifera* didn't get to China until 6,000 years after that pot was made, so the species would have to have been local. But Jiahu also has a lot of hawthorn, a red, baseball-sized, segmented tree fruit that tastes like a chalky apple and carries four times as much tartrate as grapes. Researchers there found hawthorn seeds dating back to about the same time.

Rice, honey, wild Chinese grapes, and hawthorn — it would have been hard to mix those ingredients and *keep it* from fermenting. Rice is the basis for alcoholic beverages throughout Asia, and it's in some American pilsner-style beers. Honey is the basis of mead. Fruits are the key ingredient in wines and brandies. In other words, the stuff they were pouring at Jiahu was a mixture of everything to come, the Mother Eve of booze.

This is science; people disagree. In 2012, an archaeologist named Oliver Dietrich published a paper saying he had even older evidence of human fermentation. At a site in Turkey called Göbekli Tepe, Dietrich found kitchen-like rooms with huge vats and residue that suggests malt or barley, pushing the inception date back to about 11,000 years ago. McGovern calls the results "suggestive," but absent chemical or botanical evidence, he's sticking to Jiahu.

Fermentation happens without us. It's natural — give yeast some sugar and off they go. But some time around 10,000 years ago, human beings grabbed hold of the process, tuning it to our liking and time-shifting it to when we felt like a drink (as opposed to lucking into fermented fruit on the vine). Like learning to make ceramic pots, or shape metal, or plant crops, fermentation was an early example of human beings turning rudimentary observation into science. We didn't know how it worked, but we knew that fermentation transformed something we had into something we wanted. Modifying and perfecting the process meant we were no longer mere participants in nature —we were remodeling it. Today, fermentation is common. It's a craft. But researchers are still trying to understand it, and improve it.

When I ask McGovern if I can hold the potsherd, he looks nervous. He finds me a pair of latex gloves and, donning his own, takes the brittle, broken piece out of its Ziploc and hands it to me. I have never held anything human-made that was so old. I cuddle with it a little bit, and then pull my phone out to take a picture. It's a physical link to the beginning of the human manufacture of alcohol, a palpable connection across time.

From a square-cornered industrial building on the outskirts of San Diego, across the street from a batting cage and a John Deere dealership, a company called White Labs sells yeast. Its customers are mostly home brewers, and regional-scale commercial breweries, with a few small wineries and distilleries thrown in. White Labs also has, just behind a big metal roll-up door, a small, very nice tasting room.

The room is wood paneled, decorated with artsy blown-up photographs of laboratory equipment, and behind a long, wood-paneled bar are twenty-eight taps. None of them have the decorative handles you'd see in a typical pub; each is topped by a clear plastic cylinder with a rounded end—a test tube, basically. And the flat-screen TV mounted overhead displays not a ball game but a list of small-batch beers and the yeast strains that brewed them: WLP001 California Ale. WLP585 Belgian Saison III. WLP802 Czech Budejovice Lager. And so on.

While I'm staring at the screen, Neva Parker walks up beside me. She's the head of laboratory operations at White Labs, dark haired and precise, and she says the only thing anyone could at that point: "Do you want a beer?"

I feel obliged to say yes. For journalism.

Parker goes behind the bar, picks up four small glasses, and starts stalking up and down the line of taps. She wants to show as much variety among the yeast as she can, she says, so we start with a beer made with the WLP001, an ale strain and White's most popular product. Then she pours one made with WLP051, another ale, more of a San Francisco style. Between them she sets glasses of WLP810, a California lager yeast, and WLP860, a strain for a Munich *Helles*. It's hot outside, and a little humid. Perfect beer weather. I take a sip of the 810 and a long sniff, put the glass down. I do the same thing with the 860. They're crisp and hoppy, like a pine forest crossed with ocean foam, but honestly, I can't tell much of a difference between them.

Parker seems unfazed by my philistine palate. All four beers — in fact, all the beers on tap at White Labs — are identical except for one critical variable. They share the same barley, the same hops, same water, same temperature. What Parker hopes to show off is the difference in fermentation caused by yeast alone.

People wouldn't brew beer this way in real life. Different yeasts make better-tasting stuff at different temperatures — ales tend to be hotter, faster fermentations than lagers, for example. Ingredients vary, too — different grains, and different amounts of grains, make different styles. Even the water matters. In the iconic brewing region of Pilsen, it's soft, with very few minerals. In the equally iconic Burton upon Trent in England, it's full of sulfites, calcium, and magnesium. All those things affect the success of a fermentation not only in making ethanol, but other molecules that also contribute to flavor. And this is just beer — winemaking has its own dials to turn.

White Labs aims to focus on the yeast. "Most strains can be manipulated," Parker says. "A warmer ferment will mean more ester production. Growth rates will affect different flavor compounds." Even the sugar content in the wort will change the ultimate outcome. Anyway, that all makes for a lot of permutations. Simply playing with those starting conditions, plus different kinds of grain and hops, gets you twenty-two major styles of beer, from ales like India pale and porter to lagers like pilsner and bock.

Tiny differences in fermentation procedure can have a huge effect on the final product. White Labs once made two beers with the same wort and the same yeast—the same everything, in fact, except the temperature of the fermentation. The beers tasted completely different. The beer fermented at 66°F registered a scant 7.98 parts per million of acetaldehyde, which usually has a sort of green-apple taste, but that's below most people's threshold for being able to taste it at all. The one brewed at 75°, meanwhile, had a whopping 152.19 parts per million.

Opposite the bar at White Labs is a big window that looks in on the company's real business, a laboratory where white-coated techs culture, isolate, and analyze yeast strains. Parker's team combs through yeast archives like NCYC or the American Type Culture Collection for potential fermenting strains, cleans them up if they have to, and runs trials to test speed of fermentation, flavor profile characteristics, and flocculation. Parker sells more than 50 strains, and has around 200 on cold storage—with another 500 or so kept for customers who want a sample of their yeast stored off-site.

The product is yeast, but the promise of White Labs is control—specifically, control of fermentation. In its simplest form, the reductionist version, that starts with glucose and ends with carbon dioxide and ethyl alcohol, or ethanol.

In organic chemistry, which is what we're really doing here, all those different prefixes and suffixes stand for specific chemical structures. Ending a name with -ol means that the molecule we're talking about has a hydrogen atom attached to an oxygen atom—that's called a hydroxyl group—attached in turn to a carbon atom. Now, that carbon atom has something else attached to it, too, a "functional group." In ethanol, the functional group is an ethyl group, C_2H_5, made from ethane. Methanol, to take a related example, has CH_3, based on methane. It's kind of neat, actually, that such a subtle difference—one carbon atom, a couple of atoms of hydrogen—is the distance between getting buzzed or dying from a poison that obliterates the body's ability to process oxygen.

Ethanol is a particularly cool molecule. It's a solvent, which means lots of different molecules that don't dissolve in water will dissolve in a solution of ethanol. It's odorless, colorless, and it burns really well — a sign of a good fuel.

It's also a potent microbicide. When yeasts squirt it into their environment, it kills off local bacterial and fungal competition. Now, yeasts can actually reverse that part of their metabolism — they can slurp back up the ethanol they've made to use as an energy source, as food. It's like a car running on its own exhaust — in times of emergency, yeasts can live on their own poop.

That leads to an existential question: If ethanol can be both chemical weapon and energy source, which is most important? Why, in other words, do yeasts make ethanol? Why is there fermentation at all?

Fermentation isn't an accident, or a byproduct. It's the way yeasts convert what they eat to energy. That's metabolism, a long sequence of chemical reactions, nipping away atoms here and adding them there, attaching and detaching electrons, all in the service of creating a molecule called ATP, adenosine triphosphate. Inside a living thing, ATP is energetic currency, the stuff that keeps the lights on.

We mammals start with oxygen and glucose and end, essentially, with the waste products carbon dioxide and lactic acid. It's the same stuff that spoilage bacteria make, and the same stuff in pickles. But yeasts end up not with lactic acid but with another molecule, acetaldehyde. And they don't stop there. Yeasts string hydrogen atoms onto that, making ethanol and ATP, and then the ethanol diffuses into the environment.

Yay! We have made booze.

In fact, using a particularly fancy piece of lab equipment, you can actually see the process take place. At the Carlsberg Laboratory in Copenhagen — the scientific arm of the brewery where Emil Hansen isolated and purified lager yeast in 1883 — a soft-spoken, ponytailed chemist named Sebastian Meier runs a nuclear magnetic resonance (NMR) system that can see inside yeasts while they ferment.

It's similar to the MRI technology that physicians use for soft-tissue imaging. But Meier goes to a whole other level. He has a ten-foot-tall steel cylinder filled with super-cold liquid nitrogen and thirty-one miles of wire, which generates a field strength of 18.7 tesla — 300,000 times stronger than the magnetic field created by the earth. That's powerful enough to set atoms oscillating in a way that lets the machine distinguish among molecules. So Meier makes glucose with an isotope of carbon — which the machine can track — and then feeds it to yeast. "What most people do is measure components," Meier says. In other words, they let a biological system run its course and then look for the results. "But in a complex system, that only gets you so far. If you want to improve a cell, you want to see it in action."

Meier can't see individual atoms dancing around the way doctors can see a torn ligament. His machine just gives him data. Molecules appear and are almost immediately transformed, each one an intermediate compound between glucose and ethanol. Carbon dioxide shows up clearly and strongly. At either end of the chain he sees glucose and ethanol, but inside the guts of the yeasts are a half-dozen other recognizable molecules. Pyruvate vanishes almost immediately. Acetaldehyde appears and disappears so fast the NMR can't even pick it up.

Meier went on to compare the results of a fermentation with a brewing strain of *S. cerevisiae* against one more commonly used as a laboratory organism. Sure enough, the lab strain produced a tiny bit of pyruvate but not much else; the brewing strain had a huge carbon dioxide peak, and was way better at making ethanol. "Why is it a better brewing strain? What's making it go fast?" Meier wonders. Nobody knows. They just know it works.

But even that doesn't really answer the question. That's *how* yeasts make ethanol . . . but not *why*.

Steven Benner thinks he knows. Benner is one of the inventors of synthetic biology — DIY genetics, building new genes and genomes from scratch. In the 1970s, Benner studied enzymes. Like all proteins, enzymes are made of subunits called amino acids, assembled in a sequence and shape determined by genes. They come in long, stringy bits, snaggly parts, and various dips and cup-shaped structures where chemistry happens — enzymes can attach proteins to each other or cut them apart, among other important jobs. If you're at all aesthetically inclined, then enzymes seem like evolutionary masterpieces.

Benner and his colleagues realized that they could use those sequences of amino acids to do a kind of molecular paleontology. They could compare the modern sequence of an enzyme to others created by related genes and reverse engineer a family tree to guess what the ancestor protein looked like. It's the same thing linguists do when they look at the way all the Indo-European languages say "home" and make a really good guess about what the word for "home" was in proto-Indo-European. "The idea was, if you could resurrect ancient proteins, bring them back to life, and study them in the laboratory, you might be able to understand physical behavior," Benner says. His team called their new approach "paleogenetics."

Yeasts eat sugar, but 150 million years ago, grasses hadn't evolved yet, and sugar cane is a grass. There weren't any flowering, fruit-growing plants yet, either. And yet somehow yeasts survived just fine.

But about 50 million years later, during the Cretaceous period, fruity plants took over from pine trees, and yeast adapted to life inside fruit. The rise of those fruity plants, called angiosperms, meant genocide for any species that couldn't handle it, a planet-wide extinction event. Some dinosaurs survived — the ones that figured out how to eat fruits, nuts, and berries are what we today call birds. And some ancestors of primates managed it, too, eventually leading to us humans.

"That," Benner adds, "is a just-so story." He means, essentially, that it's a theory that fits the data, but you could come up with a bunch of other theories that are equally good. Evolutionary biology is full of them.

Benner wanted to see into the deep past, to the moment when yeasts created fermentation. His time machine was made of enzymes. Yeasts turn acetaldehyde into ethanol with an enzyme called alcohol dehydrogenase 1; they turn ethanol into acetaldehyde, reversing the process, with alcohol dehydrogenase 2. In *S. cerevisiae,* the two enzymes differ by just 24 amino acids out of 348, but other related yeast species make their own slightly different versions. Benner and his colleagues sequenced a bunch of those and compared them. Then they took what they knew about how sequences change over time to create a possible version of the ancestral enzyme, which the researchers called Adh_A.

There were too many variables to get it exactly right. In the end, Benner's group had twelve different possible versions of Adh_A. They tried them all, tossing them in with ethanol and acetaldehyde to see which direction the equation ran. The result? Adh_A was much, much better at converting acetaldehyde to ethanol than vice versa. So, problem solved, right? The ur-yeast made ethanol to kill its enemies, not to consume in a celebratory way afterward.

Sadly, no, says Benner. "What the yeasts are doing is becoming resistant to ethanol by making it the final step," he says. "You and lactobacillus have become resistant to lactic acid. It's a choice that has to do with the details of your ecosystem." Yeasts don't have a circulatory system to cycle byproducts away the way ours gets rid of lactic acid; the little guys have to rely on their environment. It seems likely that back before flowers and fruit, yeasts lived in tree exudates (sap, in other words), exposed to the air. And remember why Patrick McGovern can never find direct evidence of ethanol in any of his archaeological samples? It's the same reason that, as we'll see later, distilling works: Ethanol is highly volatile — it readily evaporates. Yeasts weren't trying to sterilize their surroundings. All they want to do is get their garbage out of the house, and packaging it as ethanol was the most efficient approach.

Then the angiosperms came around. "Now they're living in fleshy fruit," says Benner. "But they're already resistant to ethanol. So the yeast is now all of a sudden pre-evolved. You're looking at what appears, in retrospect, to be the yeast getting ready for fruits. But that's not right."

Why does yeast ferment? Because it was the smartest way to live, evolutionarily speaking, on a changing planet. "And then you're going to say, well, hell, I have all the alcohol, I might as well use it. And you're going to drink it," says Benner.

"Of course," he says, "that's a just-so story."

All those strains of yeast in the brewpub at White Labs — not to mention the ones that make wine or sake or distilled spirits, plus the wild ones — make ethanol. Yet each one makes a product that tastes different, depending on starting conditions. Pinot Grigio doesn't taste like Pinot Noir. Beer doesn't taste like wine. Clearly there's more to fermentation than spinning glucose into ethanol. The molecular route from glucose to ethanol is studded with side roads and turnoffs. Metabolism is about energy, yes, but it's also about the nutrients that make it possible to build proteins, the ingredients of the cell wall and membrane, and preparing for and undergoing reproduction.

Pasteur noticed this first—that his fermentations contained more than ethanol. He spotted glycerol, butyric acid, succinic acid, and cellulose. Careful experimenter that Pasteur was, he realized that what came out depended on which yeast he used and how he used it. It was another important piece of evidence that living cells were transforming organic compounds into entirely different organic compounds.

Some of those differences come from the things that yeast doesn't touch. The colors of wine, for example, come from pigments in the skin of the grape. The colors of beer, on the other hand, are mostly molecules called melanoidins (same root as melanin, the pigment in skin). The heat of the kiln during malting induces the Maillard reaction in the barley sugars and amino acids—just like browning in a Dutch oven. Since barleys intended for ales tend to be kilned for longer, ales tend to be darker in color.

South African by birth, Isak Pretorius is the deputy vice-chancellor (research) at Macquarie University in Sydney. But before that he was vice president of research and innovation at the University of South Australia, and before that he was head of the Australian Wine Research Institute, and in those two jobs Pretorius spent years trying to improve winemaking yeast. His lab, a collaboration of winemakers, microbiologists, and geneticists, did everything from tinkering with grapes in an experimental vineyard winery to hunting for new wild species. "Commercially there are about 230 strains on the market, and not all of them are really that different," Pretorius says. "They're all *Saccharomyces cerevisiae.* They look the same. Most of them produce the same levels of ethanol. But they produce very different flavor profiles."

An example: Researchers from Stanford University and E. & J. Gallo

Winery, which has an active laboratory program, took juice from Chardonnay grapes and fermented it with sixty-nine different strains of winemaking yeast, purchased from a bunch of different providers. They measured ethanol output, but also twenty-nine different other metabolites — various alcohols, esters, acetaldehyde, sulfur dioxide, glycerol, and so on. Now, had they used different grape strains, you would've expected some variation in output. The different amino acids in the grapes would mean different kinds of alcohols at the other end, for example. But even with just one grape juice, the researchers saw variations up to a thousandfold in how much of a given chemical the yeast made, and they all did it at different speeds, too.

Pretorius, for his part, spent a lot of time tinkering with Sauvignon Blanc grapes. They're full of chemicals called thiols. (Their distinguishing trait, chemically, is that they contain sulfur.) In grape juice, those thiols are connected to an amino acid called cysteine, which makes them nonvolatile, and as a consequence we can't smell them. "But yeast has a limited capacity to decouple thiols from cysteine," says Pretorius. "That gives you that typical Sauvignon Blanc 'passion fruit' or 'tropical fruit' flavor." The compound comes from the grape, but the yeast — as part of its metabolism of the juice — makes it volatile. The same goes for the rosy- or violet-smelling terpenes in Gewürztraminer grapes. In the juice they're bound up, nonvolatile. Yeast liberates them.

With that in mind, Pretorius tried to tune yeast to produce specific effects. From a bacterium, Pretorius's lab pulled a gene for an enzyme that breaks the connection between thiols and cysteine, and engineered it into a winemaking yeast. Then he got some Sauvignon Blanc juice from Australia's hottest regions, which would usually be pretty bland — the grapes need a cooler climate to ripen well. Pretorius's engineered yeast fermented that juice into wine with thiol levels twenty times higher than any other. "When we poured that wine out of the experimental flask in my office, within seconds people fifty meters away in the reception area smelled it. It was completely over the top, but it proved the point," he says.

Importing a specific gene — or tracking down strains that already have it, as Pretorius's lab eventually did to avoid the international stigma they'd face with a genetically engineered winemaking yeast—is still a primitive approach. Yeast researchers are starting to get a little more elegant, connecting a given gene to a specific aroma or flavor. This work has just begun; even though yeast was the first organism to have its genome sequenced, the strain the geneticists used was a research one, and the sequence they got not specific enough to focus on taste. But White Labs is working with the genome powerhouse Illumina to sequence many of their strains. It's the beginning, many researchers hope, of a program to pin down which genes do what during fermentation.

In some drinks, yeast doesn't do the job of fermentation alone. Other microbes, largely unsung, help out—though even booze researchers don't always know exactly how.

In American whisky making, it's common for distillers to use some of the fermented grain in the mash—before it goes into the still—to mix in with the next batch to be fermented. That portion is called the "backset," and the whole process is called "sour mash," just like it says on the bottles. The practice began in the 1800s, probably as a way to keep the same yeast strain consistent. But that's not the only thing you can do with fermented mash.

Rum, you would think, should be relatively easy to make. Molasses or cane juice provides a huge amount of readily available sugar for yeast to chow down on. That turns out to be less an opportunity than a problem, though. Molasses can actually be low in fermentable sugars, and while pure cane juice can go straight into a fermenter, you have to rush. The warm, humid, tropical climates where the best rum comes from are extremely microbe-friendly, and those with local bugs will infect cane juice right away. You basically have to build your distillery right next door to the cane field.

That said, rum makers have figured out how to use the weird, aggressive local microflora to their advantage. Jamaica is famous for heavy, dark rums with lots of esters, the fruity-smelling molecules made of alcohols joined to sugars. Those come from yeast, but also from local bacteria. Rum production sometimes involves taking some of that backset and letting it inoculate with bacteria to be mixed into the next batch.

Even more extreme, some rum makers use a "dunder pit," a hole in the ground into which they throw leftovers from the still after production, maybe some fruit or molasses, and sometimes lime or lye to keep down acid levels. They let it sit there. For years. And this muck—they really call it "muck"—gets added back into the still. Dunder pits look as disgusting as you'd imagine, but presumably the heat of the still kills anything deadly. The exotic acids made by the fermenting action of the bacteria stay behind, though, to mix with the alcohol in the mash and form esters you wouldn't ordinarily get.

In the late 1930s and early 1940s a rum researcher named Rafael Arroyo set out to standardize the practices of his industry—worried about competition from other rum-making countries, the Puerto Rican government built him a lab and let him loose. One of the principal questions he wanted to answer was which (if any) bacteria the rum-making process really needed.

In batch after batch, Arroyo and his colleagues changed the time and the inoculation of their mash. In the end, Arroyo said, the outcome depended on what you wanted to make. "For certain of the light, very delicately scented straight drinking rums that are so much in vogue at present, bacteria are detrimental," he wrote. That was the kind of rum Bacardi was making, mostly to mix with Coca-Cola. "Some of our best rums were produced by pure selected yeast under pure culture fermentation technique. We admit, however, that greater volume and strength in taste and aroma are obtainable through bacterial and other microbiological intervention."

Arroyo didn't recommend a dunder pit, though. He wanted to know exactly which bacteria and which strains were best suited. Wild types need not apply. They had to consume little sugar, make useful acids in the right quantities, and not make alcohol themselves.

After trying a half dozen or so, Arroyo determined that *Clostridium saccharobutyricum* made the most interesting acids. He also collected and isolated a mold that grew on coffee plants that he liked. It made a rum, he said, that smelled like apples.

Arroyo's work, seventy-five years old, remains the standard reference on rum and its associated microorganisms — which is weird, if you think about it. As far as I can tell, no one has tried to sort through the microbiome of a dunder pit or isolate species other than the ones Arroyo recommended. Rum, especially strange, dark ones full of exotic esters, is one of the most underrated things to drink. But the components of its fermentation remain the proprietary knowledge of artisans. Scientists are only now trying to get specific about which microbes are most important in various drinks. In the musts of Napa Chardonnay they find the families *Firmicutes* and *Eurotiomycetes* (that latter includes *Aspergillus* and *Penicillium* fungi). But in California's Central Coast winemaking region it's *Bacteroides, Actinobacteria, Saccharomycetes,* and *Erysiphe necator.* Back up to Sonoma and you get *B. fuckeliana* and *Proteobacteria.* And other grapes have entirely other colonizers, each contributing in ways as yet unknown to the final flavor. The researchers who parsed all that call it "microbial terroir."

Ethanol and all those other metabolites make for a compelling account of fermentation, but an incomplete one. Fermentation also makes carbon dioxide. That means bubbles, and bubbles change everything.

Bakers value yeast precisely for its ability to make carbon dioxide, which opens up the little pockets in bread that make it light and tasty. The ethanol evaporates away; bakers don't need it. Most people fermenting anything — not just booze, but the people making pickles with lactic acid bacteria, too — don't do anything to contain or control CO_2. That's why you have to be careful opening a jar of Korean pickled cabbage, kimchi; the gas will bubble out of solution so fast that it'll carry liquid along with it, and you'll end up with spicy brine everywhere.

Carbon dioxide has its own flavor, which affects the overall taste of a drink. (At high partial pressures — which is to say, when a gas contains lots of CO_2 relative to other gases — it also sets off the body's pain receptors, called "nociceptors." One trick almost every distiller I visited tried to play on me was to get me to stick my head into the vat during the final stages of fermentation, when the headspace — the volume of air above the liquid — is a cloud of CO_2. Taking a whiff is like sticking a knitting needle up your nose. Too much of it, and you can pass out and fall right into the vat. Fun!)

The gas really wants to fizz away during fermentation; some breweries capture it and then inject it back into the beer. The classic approach, though, is to add a little yeast to the finished product and seal the container. Secondary fermentation, or "conditioning," produces CO_2 and scavenges free oxygen that can make beer taste weird, but the yeast can make the beer cloudy, which people often see as an impurity.

Wine, mead, sake, and the distillates don't have much truck with carbon dioxide, though occasionally you'll get something fermented with a little effervescence. Two particular drinks, though — sparkling wine and beer — depend on CO_2 as a hallmark of their drinking experience. That gets tricky, because their relationships with bubbles are utterly different.

In a bottle, carbon dioxide is under pressure, held in by the cap or cork. At higher pressure it dissolves, so you don't see any bubbles. Lower the pressure by popping the top and the CO_2 comes out of solution. In sparkling wines like champagne or prosecco, little bubbles pull fatty acids and other aromatic chemicals from the liquid to the surface. When they hit the top, they pop — a hole opens at the top of the bubble, its edges expanding at 22 miles per hour and converting to a ring of high pressure that smacks into a low-pressure region at the bottom of the bubble. The collision squirts a conical jet into the headspace of the glass that improves (or at least accelerates) the wine's perceived aroma. Also, the little splashes tickle.

A bottle of beer has 5 grams of CO_2 per liter of liquid. A bottle of champagne has 12 grams per liter. When you open one of those bottles, the CO_2 inside becomes "super-saturated" — that is, its pressure dissolved in the liquid is higher than equilibrium with the external atmosphere will allow. The CO_2 has to come out, which it does by forming bubbles. Now, champagne is pressurized to six times the atmospheric pressure on earth at sea level, enough to propel a popped champagne cork faster than 30 miles an hour. Lesson: letting the cork shoot out of a bottle when you open it is both tacky *and* dangerous.

Ideally, a glass of sparkling wine will form bubbles more subtly than the geyser from the neck of a bottle. For that to work, gas molecules have to find each other in the glass amid all those molecules of liquid. The problem is, the liquid molecules stick together. The CO_2 is like the lovers at the end of a romantic comedy, and the liquid is the crowd at the airport they have to fight their way through about ten minutes before the end.

Of course they find each other; you've seen those movies. But CO_2 molecules are actually smarter than movie stars, because the molecules have a prearranged meeting place: anywhere there's a hole of a certain size. In champagne, the cavities are on the sides of the glass, and the size is larger than 0.2 microns. The process of bubble formation is called nucleation, and in 2002 a physicist named Gérard Liger-Belair, at the University of Reims in France, set up a rig to see it happen. He got a camera capable of resolving objects as small as a micrometer — a millionth of a meter — at 3,000 frames per second, and started pointing it at glasses of champagne.

Most people expected the bubbles to come from scratches in the glass. And yes, some glassmakers laser-etch tiny nucleation sites on the bottom of their products to get visually pleasing, reliable bubble production. But Liger-Belair found something else: cellulose. Bits of cloth or paper, so small as to be invisible to the naked eye, were clinging to the inside of his champagne flutes. He speculated they were remnants of hand-drying with a towel after washing. And the inside of these cellulose fibers had just enough air space for nucleation. In fact, they became "bubble guns," launching streams of as many as thirty bubbles per second toward the surface. Champagne served in the shallow, wide glass called a coupe loses CO_2 faster than a tall, narrow flute does, by the way, so if you want your bubbly to stay bubbly, use a flute.

Beer, though, is different. The bubbles in sparkling wine hit the surface and pop, but in beer, they stick around, forming a head of foam.

In studies, people assume beer with a thick head and substantial lacing (the tracery of bubbles left on the inside of a glass) is going to taste better than beer without it.

Explaining bubbles in beer requires a conversation with Charlie Bamforth. He's the Anheuser-Busch Endowed Professor of Brewing Science at UC Davis — quite possibly the single greatest title in all of academia. Having a beer with Charlie Bamforth is like listening to music with David Bowie. At a table in a nearly empty brewpub one afternoon, a few minutes from Bamforth's lab at UC Davis, I ask him: Can you, at this point in your career, just sit and enjoy a cold brewski?

"I'm critical," Bamforth says, tilting his head in a way that suggests he's admitting a small character defect. "I think I know what I'm looking for and what should be there, and a lot of that is the prejudice of having been in the brewing industry." He'll allow for individual preference, he says. A lot of imported beer is oxidized or spoiled by the time a US drinker takes a swallow. People still love it. And that's OK. Mostly.

What's not OK? Not getting a frothy head on top. "Look at this beer," Bamforth says, gesturing at the half-empty glass next to him. "I mean, look at it. It's pathetic. The presentation's appalling." I look at his glass, and a universe of criticism unfolds before my eyes. There's no foam left at all, just a feeble ring of what looks like spittle at the edges. The lacing is nowhere to be seen. It's a freaking disaster, this beer. How could I not have seen it? "Probably 70 percent of the people here wouldn't give a rat's ass about it," Bamforth says. "If this was Germany or Belgium, that would be sent back. And I'm not doing it, am I?"

He's not. But I can tell he's having a hard time resisting the impulse.

As soon as beer hits the inside of a glass it starts to froth; by the time the glass has even a little liquid in it, beer is building a head of foam. Over a few minutes, the foam subsides. It seems simple enough, but that rise-and-fall is actually the macroscopic expression of a fierce battle on the microscopic scale. Physics wants to pop those bubbles; chemistry wants to hold them together.

On their way up a glass of beer, bubbles attract glycoproteins—molecules made of proteins and sugar. In the 1970s Bamforth and a colleague discovered that in beer foam, a glycoprotein's backbone orients toward the gas inside the bubble, and one end points outward into the liquid. Like detergents, these molecules are "surfactants"—in laundry, they get between water and dirt, pulling stains off of clothes. In beer, surfactants form a membrane around bubbles that keeps the bubbles from popping and sticks to bubbles nearby. That's why beer foam persists where champagne bubbles dissipate. Foam is a collaboration among bubbles.

The bubbles lift liquid—beer—into the foam, but gravity pulls the beer back down into the glass, most of it in the first minute after formation. The bubbles start to merge, enlarge, and pop. This is why the head eventually disappears.

Good Irish bars use another trick when they "build" a Guinness. They pull an almost-full pint from the tap, and let it sit for three minutes. Then they finish filling the glass and serve it up. It's called a "double dispense," and it lets the liquid from the second pour slip downward between the bubbles of the first layer of foam, pushing the head to the top where it seals the new foam away from the atmosphere, slowing the outward diffusion of CO_2. The stratum at the top is now stiff enough for a bartender to carve a shamrock in it, if that's your thing.

Colder beer keeps its head longer; beer poured from higher up picks up nitrogen from the atmosphere, which makes for a more stable head. These seem like prosaic solutions to the crisis in beer foam. Honestly, Bamforth seems a little dismayed by that simplicity as well. You'd think anyone could do it right. But they don't. "I'm going to try and get away from bloody bubbles," he says, moving to finish his beer at last. He looks at the un-frothy, un-laced glass. "It's just not washed properly," Bamforth says sadly. "Or they're washing these glasses alongside food plates. I don't know what they're doing, but they're not doing it right, because I can tell you the beer, when they brewed it back there, had plenty of foam potential. Ninety-five, 98 percent of foam problems have nothing to do with the beer. It's everything to do with the way it's bloody poured."

After Patrick McGovern carefully returns the Jiahu potsherd to its plastic bag and places it back on his bookshelf, we seek out lunch and beer. We walk a couple of blocks to a restaurant McGovern says will have a bottle of Midas Touch, a beer from Delaware microbrew darling Dogfish Head based on the components of residue McGovern found in a 2,700-year-old tomb. Dogfish Head makes a bunch of brews with ingredients based on McGovern's research. For one based on an Egyptian recipe, the brewers even captured wild yeast from an Egyptian date farm.

The restaurant turns out not to have any of the Dogfish beers McGovern helped develop. Neither does the one next door. Standing on the street corner, I call a couple of other joints McGovern remembers carrying it, including a tony place on a boat moored on the Schuylkill River. They're out, too. Finally, we stuff ourselves into a taxi headed for McGovern's old standby. "Sixteenth and Spruce," he tells the cabbie. "Monk's Café."

Monk's Café turns out to be a long, narrow Belgian pub that predates the upscaling of Philadelphia's Center City neighborhood. McGovern threads his way through the closely packed tables toward a back-bar dining room. The bartender waves at McGovern warmly. "This is the place where I had my very first beer that reminded me of wine," McGovern says when we sit down. It was an aged Chimay, he says, a classic Belgian ale.

This place doesn't have Midas Touch either (come on!), but the bartender tells us he has another Dogfish Head brew called Theobroma. It'll do. McGovern orders us a bottle. "It's based on our analysis of very early chocolate from Honduras," he says as the foam settles over pitch-dark liquid. "Originally they had the chocolate tree with its fruit. The beans are surrounded by a pulp with 15 percent sugar, and that has to be fermented away to get to the bean. In the process you get a 7- or 8-percent-alcohol beverage. We think that may be why people got interested in domesticating chocolate, because it made this elite beverage."

I take a sip, and it's everything a chocolate egg cream should be. And then heat blossoms at the back of my throat. McGovern takes a big swallow of his and says, "It also has ancho chiles."

McGovern admits that the later Mayans and Aztecs who made the chocolate beverage on which he and Sam Calagione, the founder of Dogfish Head, modeled Theobroma might have let the alcohol dissipate before they drank it. History suggests that no human *ever* lets alcohol just dissipate, but still, it's possible. "The trouble with these re-creations is we don't know how close they are to reality," McGovern says, powering through a bowl of steamed mussels. "We were just trying to come up with something that's drinkable. I've always said we should put more chocolate in it, but Sam wanted it understated." To McGovern's even greater regret, all the Dogfish Head beers contain barley, per US regulations. That includes Jiahu, the one based on the Chinese find, even though the Chinese didn't have barley until 3000 BC at the earliest.

Their collaboration has been productive nevertheless. Brews like Chateau Jiahu and Theobroma (plus the Egyptian-inspired Ta Henket) get good reviews and mesh with Dogfish Head's overall tendency toward—depending on how you look at it—experimentation or weirdness. In any case, the compromises between historical evidence and Calagione's instincts for modern brewing fit with McGovern's approach to research, which he calls "experimental archaeology." He can only learn so much about fermentation from artifacts and residue. These beers might not perfectly replicate what Egyptians or Mayans or Minoans drank, but at a certain point, if McGovern really wanted to commune with the brewer who first poured something into that pot in Jiahu, he had to become a brewer, too.

Four

DISTILLATION

NOW THAT MOST big-city bars have a dozen bottles of single malt whisky on the shelf, and a decent one will run you $100 in a good liquor store, it's hard to remember that single malts used to be rare. In the 1980s, the Scotch whisky business took the distinct products of individual distilleries — that's what a "single malt" is, basically — and mixed them together into blends. Johnnie Walker, Chivas Regal, Cutty Sark — that's where the money was and, to a great extent, still is. Pretty much the only single malts you used to see in the United States were Glenlivet and Glenfiddich.

That changed when, thanks to the whims of fashion and some smart marketing on the part of the big drinks companies, single malt whisky became a premium product. Demand for single malts has been rising ever since. Glenlivet, too, got caught in the rising tide. When the conglomerate Pernod Ricard bought the distillery in 2001, Glenlivet was producing and selling 275,000 cases a year worldwide — that's 650,000 gallons. A decade later the distillery sold 1.7 million gallons. Every four seconds, someone on earth buys a bottle of Glenlivet.

But no matter how much Glenlivet the company sells (and that really is quite a lot of whisky), it all has to taste like Glenlivet. With some exceptions—small bottlings and among the high-end connoisseur community—spirits drinkers aren't looking for variability in their usual. Unlike wine, with its vintages and year-on-year variation both within and among individual wineries, a distillate is supposed to be stable. In blended whiskies, that's a manageable problem. The people who make those generally start with grain spirit—and mix from a palette of single malts to match a reference sample, an actual or remembered Platonic ur-dram. The contributor single malts might change, but a master blender can adjust the quantities to get the same result. They have to. If the next bottle of Dewar's doesn't taste like the last bottle of Dewar's, you won't buy it again. The same is true for any big drinks brand, from Coors to Coca-Cola.

But what about those single malts? They can't vary much from bottle to bottle, either, or fans will bail out. To maintain that kind of quality control, a large-scale commercial distillery like Glenlivet (or Jim Beam or, well, pick your label) relies on the same principles as Toyota or Microsoft: repetition and standardization. A distillery is a factory, an industrial processing plant for fermenting grain and distilling the result into whisky.

In 2009, to meet burgeoning demand, Glenlivet built a fancy new stillhouse. It's the size of an aircraft hangar, and with its giant windows letting in the sunlight to bounce off the burnished copper stills, metal rafters painted a cheery yellow, it looks like a good place for a wedding reception. The stills are the same sizes and shapes as the older ones in the stillhouse next door, built in 1887. They're steam heated now instead of coal fired, and since 1965 the distillery has trucked in its malted barley. The yeast comes from specialist suppliers.

New facilities always come with new risk, though. Still makers keep the specs for their clients on file, because the exact shape of a still gives the spirit its characteristic flavor. But even new stills shaped like old ones don't always produce the same product. In 1990, William Grant & Sons built a small distillery called Kininvie down the road from the much larger, more famous Balvenie. The company was chasing demand—it wanted to keep selling Balvenie as a single malt but also continue to have Balvenie-like flavors in the various Grants blends. Kininvie used the same barley as Balvenie, the same water, and had stills built to the same specifications. The final new-make spirit tasted almost nothing like Balvenie; Kininvie closed in 2010. Why'd it taste different? Different microclimate at the new site? Different microbes in the air during the fermentation of the mash? Well, if they knew that, it would have tasted like Balvenie.

In its new stillhouse, Glenlivet left nothing to chance. The classic, onion-shaped stills come with a little bit of advanced technology attached—little blue packages of telemetry-gathering sensors for taking temperature and pressure readings, wired to each other and to the curving, front-desk-looking workstation where three flat-screen monitors keep track of the proceedings. "It's not the old picturesque thing," says brand ambassador Ian Logan as we stare at the monitors, "but if you were given a blank piece of paper this is what you'd build."

All six stills have little windows built into them, angled toward the desk. But nobody really needs them. The screens tell you everything you need to know. "You can not only run both stillhouses from here, but also the grains plant," Logan says. The whole place runs with far fewer staff than a classic distillery. "It may not be the romantic side that people want to see, but it's the practical side. The only way you sell a brand successfully is consistency."

Once I notice the sensor packages, I see them everywhere. They're attached to the mash tuns where the malt mixes with warm water, to the washbacks where it ferments. The stillhouse has an old-style spirit safe, a copper box into which the freshly condensed new-make spirit bubbles up, clear and bright, before it drains back out into steel holding tanks. In Scotland, spirit safes used to be locked; only the Queen's revenuer had the key, because the thermometer and hydrometer that measured alcohol content were inside the safe, and that's what they based taxation on. Typically a spirit safe also has a big lever on top, to switch the spigot over when the good stuff starts flowing from the still, and switch it back when you start getting flavors you don't want to save. Clunking a hammer-sized metal bar from one stop to another was a wonderfully mechanical, analog way to know when you were actually making whisky.

But the spirit safe at Glenlivet doesn't have a lever — the switch is automated. On the back — facing away from where the tour groups might see it — is another instrumentation carbuncle, measuring temperature, pressure, and specific gravity to a greater accuracy than the old mercury-filled glass tubes ever could. In fact, the safe is just for show. The people who run the distillery hate it. Tourists always want to touch it, and, Logan says, "It's a pain in the ass to polish."

"Do other distilleries use all this instrumentation?" I ask.

Logan gets cagey. He doesn't want to talk about other companies. But Chivas, he'll allow, relies on it. "I'm at the stage, why lie about it?" he says.

One of the stillmen — Frankie, dressed in a green Glenlivet polo shirt — takes a seat at the desk. After half a lifetime working at more traditionally operated distilleries, he's happy, he says, to be at a place run by robot overlords. That switch, from the part of the distillate you keep to the stuff you don't, has to happen at the exact right moment in the process or you ruin the batch. "I've worked on manual sites," Frankie says. "You do miss the cut now and again." And when you sell a bottle every four seconds, missing the cut would not be cool.

All the technology changes the room in one huge way: It doesn't smell like a distillery. No bread aromas, no scent of nail-polish remover or vanilla spice. Nothing. It smells like an office.

Distillers, especially whisky distillers, tell a story of craft and tradition, hundreds of years of practice that defines a culture. But the distilled spirits business is dominated by giant producers who run immensely productive facilities that produce complex, expensive chemical admixtures year after year. That's not necessarily a criticism: just because Jack Daniel's comes from a chemical plant doesn't mean it isn't a damn-fine-tasting chemical.

In fact, the process of making that chemical connects to 2,000 years of human history. Even the metaphor is potent. We distill ideas from something diffuse and hard to grasp into something precise. We distill knowledge to its essence the same way we distill fruit wine to brandy, beer to whisky, fermented sugar cane juice to rum. The writer Primo Levi called the series of phase changes built into distillation — liquid to gas to liquid — magical transformations in pursuit of the fundamental spirit of life. Distillation tells us that having *less* of something can make it *more* potent. It is concentration. It is focus.

Fermentation is a natural process, as close to a miracle as a science-minded type like me would ever acknowledge. Over human history we've learned to harness and adapt it. We domesticated the microorganisms that make it possible, designed containers friendlier to it, created businesses around it. But a winemaker taking credit for fermentation is like a beekeeper taking credit for honey. Fermentation would happen whether men and women were here on earth or not. If a fig spontaneously ferments in the forest, a monkey is there to hear it. (And eat the fig. And get drunk.)

Distillation, though, is *technology.* Human beings invented it; we came up with the process and developed the equipment. It requires the ability to boil a liquid and reliably collect the resulting vapors, which sounds simple. But to do it, you have to learn a lot of other skills first. You have to be able to control fire, work metal, heat things and cool them, make airtight, pressurized vessels. You need a big brain with a wrinkled cortex, maybe some opposable thumbs. But most of all you need a desire to change your environment instead of just live with what you have. Distillation takes intelligence and will. To distill, literally or metaphorically, requires the hubris to believe you can change the world.

If we decide to believe that William Faulkner said that civilization is distillation, when does Faulknerian civilization begin? Contemporary accounts describe still-like devices in China around 3000 BC, but in his masterful, unbelievably long *Science and Civilisation in China,* the historian H. T. Huang says the earliest record of anyone talking about actual distilled spirits in China comes from around 980 AD, in Su Tung-Pho's *On the Mutual Response of Things According to Their Categories:* "When the wine catches fire, smother it with a piece of blue cloth." Huang makes the very good point that wine with a high-enough alcohol concentration to catch fire must have been distilled — fermentation alone never gets you much above 15 percent.

We can push the origin date back even farther. Supposedly the Greek genius Aristotle, who died in 322 BC, wrote in his *Meteorology* that sailors made seawater drinkable by boiling it in covered vessels and collecting the condensate from the inside of the lid. That would be a rudimentary kind of distilling, and in fact *destillare* is Latin for "to drip off" or "to drop off." If Aristotle really wrote that, it would push the origin of distilling to the ancient Greeks. But *Meteorology* is a lost text; all we know of what it contained comes from a commentary written almost 600 years later. Pliny the Elder, who died in 79 AD, describes what sounds an awful lot like a still, so that puts us plausibly in Roman times.

For another tantalizing possibility, we could go back to what's now Pakistan, where in 1951 an archaeologist named Sir John Marshall discovered ceramic vessels he thought were used to heat and condense water. But decades later the British anthropologist Raymond Allchin suggested that Marshall had actually found stills. And a few years after Marshall, archaeologists working at a site called Shaikhan Dheri found a still, drinking vessels, and containers marked in a way that suggested they were used to hold liquor — and dated it all to between 150 BC and 400 AD. Allchin pointed out that ancient Indian literature is full of oblique references associating alcohol with the image of an elephant's trunk, and later indigenous Indian stills look elephantine, with a big pot as the head and a downward-angled proboscine projection for a spout. Based on all that, Allchin wrote, "India appears on present evidence to have been the first culture to exploit widespread distillation of alcohol for human consumption."

Yet Allchin's version isn't the accepted history. Credit for the still goes instead to ancient Egypt, to an alchemist named Maria Hebraea, more commonly called Maria the Jewess. She was a scientist from the scholarly city of Alexandria sometime between the first and third centuries. And it's . . . possible. Even though the history is sketchy, Alexandria was the kind of town where a Jewish woman researcher could have invented one of the most important pieces of lab equipment in history.

Founded and planned by Alexander the Great in 331 BC, Alexandria became a center for research and education. Alexander had been educated by Aristotle, in turn a student of Plato, who'd been educated by Socrates himself, and back in Athens Aristotle had built the first library and first museum on earth. Alexander even traveled with a copy of *The Iliad,* stored in a golden box he stole from Darius III, the leader of the Persians. So the city's first two rulers — Ptolemy I Soter and Ptolemy II Philadelphus — cemented their power after Alexander's death by hearkening back to his scholarly tendencies.

The city was on several international trade routes, which made possible the collection and copying of hundreds of thousands of scrolls from around the world. While nobody knows the exact location of the famous library, the overall layout of the city was a model of early modernity, with the first gridiron street plan — north-south avenues intersected by east-west streets, taking advantage of prevailing winds to keep cool.

Over the next few centuries, all the great minds would come to Alexandria to teach and study: Euclid, Archimedes, Galen. Two millennia ago Alexandrian researchers knew that blood circulated through the body and that the earth orbited the sun, and they named the smallest indivisible pieces of matter in the universe "atoms." Alexandria was "the single place on earth where all the knowledge in the entire world was gathered together — every great play and poem, every book of physics and philosophy, the key to understanding . . . simply everything," as the historians Justin Pollard and Howard Reid wrote. "Most of the knowledge of the first thousand years of Western civilization is missing. These were the books that formed the library of Alexandria."

On a hill in the southwestern quarter of the city was a temple called the Serapeum, headquarters for a local cult religion that like the city itself merged elements of Egyptian and Greek religion. The rooms inside the temple's double-colonnaded perimeter — or perhaps secret underground chambers — were the site of a "daughter" library of Alexandria, a backup collection. After Julius Caesar set fire to the harbor, burning

400,000 book scrolls in a battle against one of the enemies of his lover Cleopatra — it was that kind of town — the Serapeum became increasingly important as a home to Alexandrian scholarship. It was also home to a giant statue of the god Serapis positioned so that, on a certain day of the year, a beam of light would land right on his lips as though he were getting a kiss from the sun. The temple also had a mechanical "sunrise," driven by magnets. Alexandrians loved these kinds of sophisticated automata, using them in temples and parades as both common decorations and to conceal and perpetuate religious mysteries.

A man named Hero was the best-known maker of the devices. His clockwork designs, driven by heat, pressure, gravity, and magnets, included temple doors that opened only when a fire was lit in front of them, a holy water fountain that flowed when a supplicant dropped a coin into it, an orrery that moved by itself, a trumpet-playing robot, mechanical birds that really sang, and a theater where everything — sets, actors, curtains, and special effects — was automated. Hero also invented a device called an aeolipile, two copper tubes leading up from a sealed container and into a freely rotating sphere. The sphere had outlets on two sides; boil water in the lower container and steam would shoot out of the outlets, making the sphere spin. The aeolipile was, before anyone knew what to do with such a thing, a steam engine.

The point is, Alexandrians knew how to solder copper and play with steam and heat. Alexandria wasn't just a city of scientists and philosophers. It was a city of engineers — the kind of place that could plausibly invent the still.

As for Maria, historians know about her only indirectly, through the writing of Zosimus the Panopolitan, who lived around 300 AD and whose work comprises the oldest alchemical texts to survive that time. Maria is one of the two alchemists he quotes most often. Zosimus probably lived in Alexandria, but he never says where (or when) Maria lived. He doesn't describe what she looked like, or say whether she was a lecturer or researcher at the library. All Zosimus says is that she was the first of the ancients. That, and the fact that he says that the book in which Maria describes her still, *On Furnaces and Appara-*

tuses, was written by "the ancients," leads the historian Raphael Patai to conclude that Maria lived at least two generations before Zosimus. So . . . early 200s?

It's not impossible. Women researchers worked in Alexandria, a tradition that might have carried over from Mesopotamia, where women chemists developed various cosmetics. And she could have been Jewish; Hellenized Jews in Alexandria produced the first Greek translation of the Old Testament, and the city had multiple synagogues. According to Patai's *Jewish Alchemists,* Maria's writings (via Zosimus) have a distinctly Jewish cast. She refers to Jews as the chosen people, and she warns readers against touching the "philosopher's stone," the magical MacGuffin alchemists were always chasing, with their bare hands, because "you are not of the race of Abraham." Also, throughout her work she refers to "God," singular, instead of taking the more pantheistic tone one might expect of a Greek or a Ptolemaic Egyptian.

Maybe Zosimus made Maria up. Maybe Maria was real, but was describing something she learned from some other Alexandrian scientist, or from an ancient Egyptian text. In favor of an Alexandrian origin is the fact that all the other places with evidence of early distilling—India, China, Russia—are on trade routes leading out of the city. Plus, distillation is a technology that leverages the fundamental properties of physics and chemistry to seemingly transmute one material into another. That makes it something that the alchemists could plausibly have investigated. The rest is mystery.

But let's take Zosimus at face value. His descriptions of Maria's mysticism and philosophy were compelling enough to convince early Christians she was a prophet and get her written about by Carl Jung. But more importantly, Zosimus establishes Maria as an experimentalist who built her own laboratory equipment. Famously, she invented a piece of gear that still bears her name: the *balneum mariae*—in German it's a *Marienbad* and in French it's a *bain-marie,* essentially a double boiler. Heating materials to extract their essence was one of the basic moves in the alchemists' playbook; they thought they were replicating what took place within Mother Earth to turn "base metals" into the higher metals of gold and, eventually, the philosopher's stone. The *bain-marie* gave them much better control over the process. Some historians argue that the apparatus itself had been in use for centuries before Maria, and that its invention is ascribed to her through luck and historical accident. But even if they're right, it's been called the *bain-marie* for eighteen centuries. That's pretty good billing.

The second piece of equipment, the *kerotakis,* is where Maria gets to distilling. Used for heating metals and mixing them with pigments — part of the attempt to transmute base metals — it was, essentially, a heated cup into which chemicals were placed, with a tray suspended above containing whatever the alchemist was trying to alter. Maria put them both in a nearly airtight container that allowed the vapors from the cup to condense and drip down onto the target. Very still-like.

Most important to us, though, is the *tribikos.* It was a round vessel with a sort of stovepipe that had three downward-pointing pipes at the top, each terminating in a glass flask. Heat the vessel at the bottom and the material within it — for Maria's purposes, that was usually sulfur — would vaporize, rise to the top, and then condense along the pipes. Maria specified copper for the tubing, and recommended soldering the metal and using flour paste to seal the join between the downpipes and the glass flasks. The condensing element, at the top, was called *al-anbiq,* or "alembic," a word still used to describe a specific kind of still. (In today's English, it's a safe bet to assume words that begin with "al-" derive, etymologically, from Arabic — algebra, for example. More germane here, later Egyptians used Maria's equipment to make cosmetics and other powders, which they referred to as *al-kohl,* from which we get "alcohol.")

What Maria had invented fit right in with the central project of alchemy. Distillation was — and remains — a technology for separating what seems inseparable. Philosophically, Maria thought that the universe was made of four material elements (earth, air, fire, and water), four metals — what she called the *tetrasomia* (copper, iron, lead, and zinc) — and four basic pigments (white, black, yellow, and red). Wrong, wrong, and wrong, but you can see the beginnings of the reductionist thinking that's one of the engines of the modern scientific method. We take stuff apart to figure out the basic components—just like Maria wanted to. Distilling turned out to be a profound tool for that separation, right up there with filtration and gravity. But unlike those other two, which rely on the size or weight of the thing you're trying to separate, distillation uses volatility, the tendency of a substance to vaporize from a solid or liquid into a gas. This was a new way of thinking about the universe.

If you buy this account, then Maria's design spread along Alexandrian trade routes—to India, to China, to the Middle East. It got passed along, refined, reused. And that was lucky, because back home, the culture of scholarship and innovation that defined Alexandria eventually died away. In 298 AD the Roman emperor Diocletian put down an Alexandrian rebellion against Rome, and built a massive column inside the Serapeum to commemorate his victory. Four years later Diocletian was back, this time to burn all the Christian books in the libraries. He burned the Egyptian chemistry books, too. About ninety years after that, the eastern Romans destroyed all the pagan temples in their empire, including the Serapeum. Archbishop Theophilus (Greek for "lover of thinking," ironically) turned the Serapeum into a church.

A new library now stands on the Alexandria waterfront. The city is a hub of ultraconservative, fundamentalist Islam. On a low hill in the middle of the city, a broad, ninety-foot-tall red granite column with a Corinthian top—the column Diocletian built—marks the location of the Serapeum. It's the only thing left of Maria's city.

The Alexandrian alchemists don't seem to have used their still to make booze—though it seems unlikely that one of them wouldn't have at least tried it, doesn't it? The first centuries AD were a high point of wine production by Roman military veterans in southeastern France, and they exported this wine throughout the empire. The amphorae that carried it have been found in Egypt, so it's possible Alexandrian distillers had access to French wine. If you worked in the lab that invented the still, you'd at least run some wine through it after hours, wouldn't you? Just to see what would happen?

Yet no archaeologist has ever found evidence of Ptolemaic brandy.

In fact, you don't get signs of distilling to drink until sometime between 950 and 1100 AD, in — where else? — Russia. It's vodka, "bread wine," suggesting that it might have been distilled from a traditional Russian fermented drink called *kvass,* made from bread or just about anything cheap and sugary. Somewhere between 1130 and 1160 a physician named Salernus mentions the use of a still to produce alcohol for use as medicine, and in the 1200s book *De secretis mulierum* the philosopher/priest/magician Albertus Magnus — a big shot among those Dark Ages alchemists I mentioned earlier — has two recipes for drinkable distillates. He called the stuff "fire water," or *aqua ardens.* Etymologically the concept lives on in a particular class of high-proof unaged rums from South America, like the cachaça used in cocktails like Mojitos and Caipirinhas. The liquors are called "aguardentes" or "aguardientes" — literally translated, that's "firewater," and going by taste they're aptly named.

The big breakthrough, though, didn't come until the mid-1280s, when a Bologna physician named Taddeo Alderotti published a book called *Consilia medicinalia.* In it he copied a process for making a distillate he called *aqua vita,* the water of life. Word began to spread. Whatever this liquid was, it seemed to cure diseases, relieve pain, fix bad breath, purify spoiled wine, preserve meat, and draw out the essences of plants.

The neatest part was in Alderotti's advance of the technology. He added a long, serpentine outlet from the stillhead immersed in cold water to aid condensation, says the food historian C. Anne Wilson. It worked faster and increased production. The Bologna physicians were making booze.

Without knowing it, they were taking advantage of some fundamental physics. The molecules of a liquid — water, let's say — are always in motion. Occasionally one of those molecules acquires enough energy to break free of the tension holding the surface together and burst into the air above the liquid. Add energy to the liquid in the form of heat, and molecules achieve escape velocity more often. In other words: heat a liquid and it evaporates.

Now, as a thought experiment, imagine that happening in an airtight container, in a vacuum. The vapor coming off the liquid pushes on the walls of the container, like any gas. That force is called vapor pressure, and it goes up or down in proportion to temperature. In other words: heat a gas, and it expands.

Dump enough energy into the liquid and the vapor pressure equals the surrounding pressure. That's the boiling point, when the liquid phase changes to gas. (Bubbles in champagne or beer, you might remember from the last chapter, are carbon dioxide; bubbles in a boiling liquid are gas that was once the liquid itself.) That's why cooking instructions change if you're at altitude, like on a mountaintop, where the ambient pressure is lower. It's also why, if you were exposed to the hard vacuum of space, you wouldn't explode. The liquid in your mouth and on the surface of your eyes would just boil away, right before you froze.

Water, for example, has a boiling point of 212°F. It has a lower vapor pressure than ethanol, which boils at 173°F. Ethanol is the more volatile molecule. This is the key to distilling booze. The interesting stuff vaporizes and comes out; the water stays behind.

Because ethanol is more volatile than water at every temperature, distilling will eventually pull all the ethanol out of a solution—to a point. At 95.57 percent alcohol, the vapor has the same ethanol concentration as the liquid, and you can't squeeze any more out. It's called the azeotropic limit, and it's the upper bound on how potent any spirit can ever be: 194.4 proof. ("Proof" is an old word when it comes to alcohol content; in the United States it's just twice the percent alcohol-by-volume. "Eighty proof" is 40 percent ABV. In the United Kingdom it has a slightly different calculus—100 proof is 57.15 percent ABV, a definition that comes from the days when British navy rations included rum for sailors. The sailors would mix gunpowder into the rum, and if it ignited when lit, that was proof the alcohol content was high enough to serve.)

You don't have to know any of this to operate a still. The medieval alchemists didn't, and they pursued experiments with metals and other materials into the Renaissance. Physicians didn't, but as of the 1200s they used various distillates of wine and herbs medicinally. (Versions of these drinks are still around—Chartreuse and Bénédictine are my two favorite examples.) Every volatile ingredient in whatever you're distilling has its own vapor pressure, and those combine in chaotic ways. Those other molecules are called congeners — they're everything in a drink other than ethanol and water, and they give distillates their flavor. (Unless you're making vodka, in which case the whole point is to pull all of those out — to leave nothing but water and ethanol behind.)

The technology spread from a few medical schools and apothecaries to monasteries all over the continent and England and Scotland, if for no other reason than people like to drink and the economics make a lot of sense. Farmers could harvest all their grain or fruit, distill it down to a few easy-to-transport barrels of liquid that would never, ever spoil, and the liquid was worth more at market than the source material. Distillation was literally a transformative technology.

When the Black Death spread across Europe, from 1347 to 1350, physicians didn't really have anything to make people feel better besides aqua vitae. So they used them. The French translated *aqua vita* into *eau de vie*, and the Dutch called it "burnt wine," or *brandewijn*. Exported to England, that got corrupted to "brandy-wine," and eventually just "brandy." The Scots started making the stuff out of grains; in Gaelic, they called it "water of life," *usquebaugh*, eventually corrupted to "whisky." By the early 1400s, people were getting addicted to ethanol. Liquor had spread across the world.

Maria's alchemists and Alderotti's physicians might recognize a modern distillery, but only barely. Scotch whisky is still a batch process, taking a single quantity of fermented mash, distilling it all, and then starting again. The stills are enormous, though they look a little like Maria's tabletop *tribikos*. As different spirits have evolved, they've taken on different setups. In Ireland, they use three pot stills to make whisky; in Scotland, it's two. Clear, light, Puerto Rican rum uses a continuous still, but heavier, darker styles of rum generally use two or three pot stills chained together. Of French brandies, Armagnac is distilled only once in a copper pot still, but cognac twice. Eau de vie is double distilled in a copper pot; aquavit comes off a continuous still. And so on.

But in places like Kentucky, the heart of the American distilling business, distillers use an even more industrialized version—and that comes with its own set of problems.

In 1996, George Shapira died at ninety-two years old, the last of four brothers who founded Heaven Hill Distillery in Bardstown, Kentucky, sixty-two years prior. That same year, the distillery caught fire. High winds spread the flames to the warehouses that held the aging spirit. Oak barrels exploded, rocketing into the air and spewing flaming whisky like napalm. A river of fire spread across the road nearby and flooded two miles of the creek that supplied water to the distillery. By the time the fire was out, the distillery, seven warehouses packed to the rafters with barrels, three trucks full of grain, and 7.7 million gallons of whisky were gone.

Three years later, in 1999, Charlie Downs and Craig Beam arrived at the Bernheim Distillery near downtown Louisville, forty miles north of Bardstown. Downs was Heaven Hill's distillery manager, the guy who keeps everything working. Beam shared the duties of master distiller with his father, Parker Beam, Heaven Hill's master distiller for thirty-nine years and counting. (Parker's father, Earl, became the first master distiller after he quit his job as assistant to his brother Carl at the family company, Jim Beam.) There'd been a distillery on the Louisville site since the late nineteenth century, and the two Beams, with Downs's help, aimed to start it up again.

The equipment was still there—grain silos, fermenters, and two six-story-high column stills. But it had never run on Heaven Hill's particular recipes of corn, rye, wheat, and barley. (Scotch whisky uses barley only; American whiskies mix in other grains, primarily corn and rye.) Heaven Hill makes a bunch of spirits, and each recipe behaves differently while it's being pumped and piped through production. The only way Heaven Hill could know what would happen to their mix as it went through the stills would be to try it.

The column-shaped still was a tremendous advance over pot stills. In a pot still, you make a batch, and then clean out the stills and start fresh. Column stills can operate continuously, a handy feature if you're trying to upscale a commercial process. In 1813, a French researcher named Jean Baptiste Cellier Blumenthal figured out that by adding little ledges into the long neck of a still you could increase the number of cycles of evaporation and condensation before the vapor flowed out. What he'd invented, working from the advances of a few others, was continuous distillation using a column. This came to be called fractional distillation (or sometimes rectification), and it let a still run twenty-four hours a day, as long as it had feedstock and steam. The old batch method separates the components of a distillation by time — into heads, heart, and tails (sometimes also called foreshots, heart, and feints). The foreshots you'd toss; the heart was the stuff you wanted to keep, or maybe redistill. And the tails could sometimes go back into another batch. A continuous column still, on the other hand, separates components not in time but in space. The stuff that comes off the top — the heads — are the lighter, more volatile molecules, and what comes off the bottom — the tails — are more concentrated and less volatile. It's roughly the same technology used to crack crude oil into gasoline, kerosene, diesel, and so on.

An Irish inventor named Aeneas Coffey perfected the system in 1830. Working at Dublin's Dock Distillery, he built a functioning, continuous, columnar still that remains to this day the basic model. Fermented mash goes in at the top and tumbles down over metal plates. Steam rushes upward through holes in the plates, picking up alcohol on the way out. Esters and aldehydes are more volatile than ethanol alone, so they distill out first, and the lightest ones aren't exactly palatable. Another thing that comes out first, before the drinkable stuff, is methanol, which is why poorly made booze can kill you — or make you go blind. Organic acids and phenolic compounds (like the taste of peat in whisky) are less volatile than ethanol, and nontoxic, but they can have a metallic flavor. They come out of the still last. (Pigments like the anthocyanins that make red grapes red are nonvolatile, so they stay back with the basic ingredients. That's why what comes out of a still is clear — "white dog," or as illicit distillers used to call it in the United States, moonshine.)

In action, the plumbing makes a Rube Goldberg device look elegant. The main still in a big bourbon distillery in the United States might be as wide and tall as an ancient sequoia, and have dozens of plates inside. And it's only the beginning of the process; spirit comes out of the column and then goes to yet another subsidiary still. You can chain several together — some rum distilleries employ as many as five columns, each one designed to pull a different fraction out of the distillate in succession.

Between the foreshots at the start and the feints at the end you get ethanol and "fusel oils," also called "higher alcohols": amyl alcohol, butyl alcohol, and so on. Depending on the ethanol concentration, they can be more or less volatile, and getting the amount of them right in the "heart," the middle part of a distillation that you might actually want to drink, is the art of the distiller. The points at which the distiller makes the cuts are sacrosanct — whether by timing what comes out of a pot still or placing the plumbing just so on a column. Other than the shape of the still, that has the most impact on the flavor of the spirit.

When Downs and the Beams first bubbled the Heaven Hill mash into the Bernheim column stills in 1999, it clogged. They had to shut everything down. "We had to go inside the still and drill some holes to make the beer flow right," Downs says. "That was a big mess." Beam just shakes his head at the memory. They had to open up the "manheads," the holes for a person to get access, and put half a dozen people inside the stills "washing and scrubbing. We lost a good day's production, and we had to fix it there and then."

Eventually they got the stills running and made their first batches of white dog. It didn't taste good. What bubbled out of the spigot at the end of all that plumbing was slightly vegetal, sulfurous. Downs knew the answer: he needed more copper.

In Scotland, stills come almost exclusively from a company called Forsyths. Diageo owns the only real competition, Abercrombie. A German company makes high-end stills for small-scale distillers in the United States. But just about everyone else in the States goes to Vendome Copper & Brass Works, a couple miles from Heaven Hill.

Though its headquarters is a small converted nineteenth-century hotel attached to a humble-looking machine shop (alongside a 50,000-square-foot warehouse), Vendome has a huge footprint. The company has built stills, mash cookers, fermenters, tanks, and piping for every big American whisky maker, from Buffalo Trace to Maker's Mark to Four Roses, not to mention fuel ethanol stills for Archer Daniels Midland and Midwest Grain. (The process for fermenting and distilling corn into ethanol is in substance much the same as the one for making whisky — but the yeasts tolerate higher ethanol levels and taste, obviously, isn't a consideration.)

Rob Sherman, the vice president of Vendome, is the fourth generation of his family to make stills. On the day I visit, he's a little harried. His wife is out of town, he has to pick up the kids from school, and tour groups from the American Distilling Institute conference have been trooping through his shop all day. Still, I cadge a look around. The work area is crowded with projects for every recognizable name in American distilling — a craft still for Jim Beam, bulldozer-sized yeast tanks for Jack Daniel's, a flute-like column eight inches wide and twenty feet high destined for R&D at Iowa State.

Sherman talks about copper the way a woodworker talks about high-quality wood. "You can anneal it, let it cool down, or work it red hot," he says. "It'll stay really soft. It rarely fatigues. Copper can outlast stainless four to one." Heat — thermal energy — is actually mechanical movement, and in a solid with springy interatomic attachments like copper, that energy creates tiny whorls called phonons that move through the medium like a wave of sound does through air. When someone says that a metal like gold, silver, or copper is a good conductor of heat, what they're really saying is that the metal is good at propagating phonons. And copper's crystal structure, the way its atoms line up in bulk, is really good for working into new shapes — its atomic crystals are smoother on their faces than in other metals, so they slide across each other, metallurgically speaking.

Of thermally conductive, workable metals, copper is the cheapest. But it also turns out to have properties critical to the flavor of distillates. Yeast metabolism makes a lot of sulfur compounds, but most of them stay inside the bodies of the yeast. Get rid of the yeast corpses at the end of fermentation, as most winemakers and brewers do, and that's no problem. But if you leave the yeast in the mash, as distillers tend to, the sulfur-bearing molecules spill into the mix as the yeast bodies crack open. You end up with hydrogen sulfide, which gives rotten eggs their odor, and dimethyl trisulfide, which tastes like rotten vegetables.

Copper, though, bonds with sulfur much better than hydrogen does. Some elements just have a type like that. Silver, for example, likewise has an affinity for sulfur, which is why it tarnishes — atmospheric sulfur, a byproduct of burning coal, sticks right on. Aluminum does the same thing with oxygen; pure aluminum used to be more expensive than gold because it was almost always trapped as an oxide. It took the development of an energy-intensive electrolytic process to make aluminum foil a kitchen-drawer mainstay.

Similarly, the universe prefers copper sulfide to hydrogen sulfide. Inside a copper still, hydrogen sulfide is going to get pulled apart, and the sulfur is going to stick to the copper, forming a patina — a sort of rust. Even with new, clean copper, what looks like a smooth, burnished surface actually has microscale peaks and valleys, and all that surface area provides reaction sites for esters and other aromatic molecules to join into new compounds.

Rice's low sulfur content means that if you're distilling rice wine — as they do to make baijiu, shochu, soju, or whatever your language calls distilled sake — you can build the still out of steel. For whisky and other brown liquors, you need copper. How much is an open question. In 2011, researchers at the Scotch Whisky Research Institute — more about them in Chapter 6 — finally did the obvious experiment. They had the coppersmiths at Forsyths build two laboratory-sized stills. One was all copper, and the other was stainless steel. (They also built a third one from copper with six components that the researchers could swap out for stainless, to see if any one was more critical than the others.) Then they used the same ingredients to make whisky in both stills, which they served to tasters and ran through a gas chromatograph. The spirit from the copper still tasted "clean," "pungent," and "cereal," reported their panel, while the stuff from the steel tasted "sulfury" and "meaty." And indeed the steel-distilled spirit contained significantly higher levels of dimethyl trisulfide and a bunch of other sulfur-containing compounds.

So, problem solved. Except when copper in the still turns into copper sulfide, it blackens and flakes off. The walls, about half an inch thick at most, get thinner and thinner. In Scotland, stills have a lifespan of just twenty-five years or so. The stillmakers at Forsyths do maintenance checks every year on the stills they built, and replace worn parts with new sections until it's time to replace the whole thing. And when they do, they replicate all its structural idiosyncrasies down to dents and dings. Short, squat stills let through heavier, less volatile molecules than tall, graceful ones; distillers think even quirks in the shape of a pot still's neck are the difference between achieving a specific flavor or missing it.

In Kentucky, Vendome is pretty much the only still-making game in town. So after their white-dog debacle, Heaven Hill called Sherman's people in for a consult. Vendome made the Bernheim stills with plenty of copper, but the Vendome team recommended adding more, in the form of a giant screen that would fit into the doubler like a coffee filter. It sees so much sulfurous action Heaven Hill has to replace it every year.

Today you can buy Heaven Hill and all its various other labels, and they taste good. But Downs and Beam still aren't exactly satisfied with their new digs. "We had more copper in Bardstown," Beam says. The white dog isn't what it once was. "It's very close, but there's some differences."

"Really?" I say. I have by now taken a couple sips of white dog fresh from the still upstairs, and it was tasty. "Like what?"

Downs cuts in hastily. "If we set the white dogs side by side, you wouldn't be able to tell the difference," he says.

"I doubt if a normal person would be able to tell the difference," Beam agrees. (I'm clearly the "normal person" at the table.) But then his affect changes — Beam's posture and face telegraph resignation. "My grandfather always said, you could take a different distillery, and even with the same recipe, you get a different whisky." Moving to Louisville saved the business Beam's family has been part of for three generations. Heaven Hill is still Heaven Hill. But because of variations in the crystal structure of copper — or some difference in the local ambient microbial population, or something even more unknowable — Heaven Hill will never be Heaven Hill again.

The scents of spearmint and anise fill the vast, sunlit interior of St. George Spirits — today, at least. The distillery, headquartered in a former aircraft hangar in Alameda, California, is in the middle of a three-week-long distillation of absinthe. Clear, herbaceous liqueur is flowing out of two-story-high, gleaming copper stills that stand in the middle of the old hangar. With their Fiat 500–sized vats and tall, porthole-lined columns, the stills look like a steampunk cityscape, something Jules Verne might have drawn on his high school notebook.

St. George combines technology that the Bologna physicians could have used with a few very modern modifications. It's a small producer (compared to a place like Heaven Hill), one of the leaders of the burgeoning craft distilling movement. American spirits are having a little bit of a renaissance — akin to what happened with small-scale brewing in the 1970s and 1980s, when brewers like Fritz Maytag at San Francisco's Anchor started making a premium product to compete with the industrial lagers like Coors. Distillers face a slightly different landscape than the beer pioneers — criticism of the products of big distilleries in the United States is more muted, for one thing. Wild Turkey makes an ocean of whisky every year, but it's a delicious ocean. Also, laws don't favor people learning the craft at home before trying to start a business. Craft brewers can learn their trade messing around in basements and garages, but eighty years ago someone who wanted to sell a homemade distillate would have to be good at distilling and also good at driving his car quickly away from government agents over back roads.

That second part is changing. Trade organizations are ramping up their lobbying to make it easier for distilleries to get licensed and sell their products. The United States even has its own university-level distilling program now. Before Prohibition the United States had more than 10,000 distilleries; as of 2012 there were just about 250 at the craft scale. But twenty years before, there were only 4.

Other West Coast stalwarts of the microdistilling scene have signature products — California's Germain-Robin is known for brandies, and Clear Creek, in Oregon, for its eaux de vie. St. George got noticed for a line of vodkas still sold under the label Hangar One, but Lance Winters, the master distiller at St. George, is a dominant figure in the world of artisanal distilling because he pushes medieval technology as far into the future as it will reach. St. George sells a suite of popular gins, a rum, a tequila, and, of course, the absinthe. Winters's small-scale experiments — eaux de vie of apricot, kombu seaweed, crab, foie gras, marijuana, and so on — are justly legendary among booze geeks. (Drinking the seaweed is like getting hit in the face by the ocean, and the apricot tastes Platonic. No kidding, his apricot eau de vie is the philosophical qualia of apricot. It is like drinking the design spec.)

Walking around St. George amid various barrels and bottles, chasing Winters past decontextualized detritus — four long brass pipes from a pipe organ? a candlestick telephone? — while he dips into casks with a glass gadget for extracting tastes (called, beautifully, a "wine thief") is one of the great ways to spend an afternoon. You're supposed to pour what you don't finish onto the concrete floor, but every time I do, I die a little.

For craft distillers a giant column like the one at Heaven Hill is cost-prohibitive and too finicky. St. George uses a hybrid still, a pot attached to a column with plates inside that you can click into or out of use, and room to suspend botanicals inside if you're making gin. Turn the right valves and it'll act like a pot still, a column, or a combination of the two.

While the absinthe runs, distiller Dave Smith is babysitting the "laboratory still," a seven-foot-tall, thirty-liter concatenation of copper and steel mounted in a tiled nook in a room off the main production floor. It's next to a wall of shelves lined with glass jugs half full of clear liquid, handwritten tags tied to their necks like something out of a pre-industrial apothecary — cinnamon, California laurel, orange peel. Boiling in the guts of the still today is another experiment: sweet potato shochu, an unaged Japanese distillate usually made from tubers or barley. (It won't be shochu exactly, because Winters used bottled enzymes to convert the starch to sugar instead of the classic koji mold.)

Smith, a compact guy with a shaved head, puts the back of his hand against the pot. It's almost too hot to touch, but the stainless-steel pipe that points upward from it and then banks down into a copper, porthole-lined column is still cool. There isn't enough energy in the system yet. St. George doesn't have the kind of sensors Glenlivet does, or the sewage-treatment-plant control rooms you'd find at a big Kentucky bourbon distillery. Winters and Smith have their hands and their palates.

As heat and steam move, hidden, through the pipes and valves next to him, Smith frets like an expectant father—looking through tiny windows, feeling valves, taking notes. "If everything goes right, anyone can do this. What experience gives you is how to handle it when things go wrong," he says. "And things always go wrong."

Smith feels the still again. Now he'd risk a burn if he rested his hand on the pot; the stainless-steel pipe is hot to the touch. Clear liquid is running from the spigot at the bottom of the condenser, next to the pot; it has a built-in hydrometer to measure the ethanol in the distillate. It clocks in at just under 80 percent—that'd be 160 proof. Smith dips a finger in and takes a whiff. (No chance of contamination in something that's 80 percent alcohol.) "I don't get any acetic acid or ethyl acetate," he says. That's a good sign. Vinegar- or nail-polish-remover-like aromas would indicate something wrong with the fermentation. I take a dip—there's something almost like cocoa powder and cake batter in there. And it's viscous, oily.

Smith notices a bead of moisture sitting at the point where the stainless-steel pipe comes out of the neck of the pot. He wipes it away with a cloth, and it reappears. It's a leak. Smith frowns. "What do you think the odds are we can fix this without blowing up?" he asks. This is not a question I can answer, so instead I back away uneasily. Smith laughs at me and fetches a wrench.

A few minutes later what's coming from the spigot is totally different—and to me, tastes deeply of potato. But the proof is dropping too fast. Smith turns a valve to raise the amount of water in the column above and the ratio shifts back toward ethanol in the distillate. "That's a little better," he says. "What I like about it is that the cocoa is still there." Pretty soon, off flavors will start showing — the tails.

Winters comes in to check Smith's progress. Like Smith, Winters has a shaved head, but Winters is a big guy who walks around chewing an unlit, tar-black cigar the size of a roll of quarters. He moves around St. George like a captain walking the deck of a ship — not totally surprising, since Winters started his career in complex, finicky plumbing working on the nuclear reactors of the aircraft carrier USS *Enterprise.*

Smith offers an update: "There's some potato. Some cocoa. Some marshmallow," he says. "The tails aren't awful."

Winters takes a sip. "That definitely speaks to doing one with at least twice as much potato content," he says. Smith pours off the distillate into an 800-milliliter Erlenmeyer flask and labels it with the words "sweet potato" and the date. He jams a stopper into the neck. The question is, will it change if it sits for a while? Winters is skeptical. "I don't know what acids are already there to esterify," he says to Smith. Winters swirls the flask, takes the stopper back out, smells it. "This totally reminds me of some of my home distillations of beer," he says.

"I was thinking of pisco," says Smith. (Pisco is unaged brandy made in South America.)

Eventually they decide to try another batch, doubling the amount of sweet potato to raise the alcohol level and increasing the heat under the pot. "That way the fatty acids in the yeast will have more alcohol to react with, so there's more potential for the creation of esters," Winters says.

He makes a few notes in a leather-bound lab notebook, with a fountain pen. As a small distiller with a reputation for innovation, St. George isn't bound to consistency. That's what the laboratory still is for — it's all in the spirit of experimentation. Smith shrugs a little, picks up the big Erlenmeyer, and stows it on a shelf.

Five

AGING

THE AIR OUTSIDE a distillery warehouse smells like witch hazel and spices, with notes of candied fruit and vanilla — warm, mellow, and tangy, the aroma of fresh cookies cooling in the kitchen while a fancy cocktail party gets out of hand in the living room. James Scott encountered that scent for the first time a decade ago in a town called Lakeshore, in Ontario. Just across the river from Detroit, Lakeshore is where barrels of Canadian Club whisky age in blocky, windowless warehouses. Scott had recently completed his PhD in mycology at the University of Toronto and launched a business (just him, a couple of employees, and a website, really) called Sporometrics. Run out of his house, it was a sort of consulting detective agency for companies that needed help dealing with fungal mysteries — figuring out why production lines were contaminated, assessing whether indoor mold was dangerous, stuff like that. The very first call he got was from a director of research at Hiram Walker Distillery named David Doyle.

Doyle had a problem. In the neighborhood surrounding the Lakeshore warehouses, homeowners were complaining about a mysterious black mold coating their houses. And the residents, following their noses, blamed the whisky. Doyle wanted to know what the mold was and whether it was the company's fault. "When he originally described it, it sounded like this black mold was growing inside people's houses and there was some psychic connection between it and the distillery," says Scott. It didn't sound worth his time. But Doyle offered to pay his way, so Scott headed down from Toronto to take a look.

When he arrived at the warehouse, the first thing Scott noticed (after "the beautiful, sweet, mellow smell of aging Canadian whisky," he says) was the black stuff. It was everywhere—on the walls of buildings, on chain-link fences, on metal street signs, as if a battalion of Dickensian chimney sweeps had careened through town. "In the back of the property, there was an old stainless-steel fermenter tank," Scott says. "It was lying on its side, and it had this fungus growing all over it. Stainless steel!" The whole point of stainless steel is that things don't grow on it.

Standing next to a black-stained fence, Doyle explained that the distillery had been trying to solve the mystery for more than a decade. Mycologists at the University of Windsor were stumped. A team from the Scotch Whisky Research Institute had taken samples and concluded it was just a thick layer of normal environmental fungi: *Aspergillus, Exophiala,* stuff like that. Ubiquitous and—maybe most important—not the distiller's fault.

Scott shook his head. "David," he said, "that's not what it is. It's something completely different."

He grabbed a few samples and took them back to the lab he was using for his Sporometrics work. (It was his kitchen table and a couple of microscopes in his guest bedrooms.) Magnified, the black stuff looked to be a mélange of fungi, but much of it was thick-walled, rough-skinned stuff he'd never seen before. It looked like chains of roughly hewn barrels. Scott realized where Doyle's other researchers had gone wrong. "They would have taken it and scraped it over a petri dish," Scott says. "And what would have grown were spores that just happened to be passively deposited there." In other words, the more common fungi would land on top of the mystery stuff, and then grow faster in culture. Come back in a couple weeks and the petri dish would be full of the boring species, but not the black sooty stuff on the walls.

Scott had a different trick. He ground up the sample and seeded a petri dish with it. But then he put the entire dish under the microscope and, using an impossibly fine needle, picked out just fragments of the rough-skinned fungus and transplanted them to their own dishes. Considering that microscope optics mean that everything you see through the eyepiece is reversed from its true position, it wasn't easy.

In the paper Scott would eventually publish about all this, he wrote that he did the transplanting with a number-00 insect pin, a steel needle 0.3 millimeters wide used to impale bugs for dissection or display. That was a lie. He actually made his own needle out of tungsten wire, a trick from the 1920s he picked up from a friend, a forensic microscopist in Chicago. (Melt some sodium nitrite powder in a test tube over an open flame, then stick in a tongue depressor and let the metal cool. When it solidifies, break the test tube — you'll have a sodium nitrite Popsicle. Soften one side over the flame and then drag a red-hot, 1-millimeter tungsten needle across the metal. It will pop and flare, and molten tungsten will spray everywhere as the wire loses diameter. Wear gloves, and put something fireproof on your countertop. Don't try this at home.) "You can etch the needle down to a 1-micron point," Scott says. "And they're super-tough needles, too."

Eventually Scott had individual spores or fragments of the mystery fungus in fifty or sixty petri dishes. He checked them every week, but found . . . not much of anything. Under a microscope the samples were clearly the same black, uneven barrels. Even after a month, his colonies were vanishingly small. Whatever the fungus was, it wasn't growing like it grew at the distillery. "I figured, there has to be an easier way."

Making growth media for fungi is really about preparing food they like to eat. So Scott went out and bought a bottle of Canadian Club. "I put maybe a shot of whisky in a liter of agar, and I poured the petri plates with that," Scott says. "That made it grow a hell of a lot faster."

The fungus clearly had an affinity for booze. But he still didn't have a clear link. Now, though, Scott was in the grip of the mystery.

The only child of a hairdresser mom and a dad who drove a backhoe, Scott grew up in a small town near Lake Erie. He was the first person in his family to go to college, but not out of any particular drive. In fact, he attended his first mycology lecture with the intention of skipping the rest of the semester; he was hoping to find someone who would lend him notes.

Then the professor told a story about a fungus that lives on peach pits. No one, he said, knows how the fungus gets from pit to pit. "If you go to an abandoned orchard and lie on your stomach under a tree for a week, watching which insects land on a peach and move to another one," Scott remembers him intoning, "you will know more about this fungus than anyone in the world."

"It was something even I, an undergraduate who didn't know anything, could do," Scott says. "I could go out there and look for stuff." In the space of one anecdote, Scott had become a mycologist. You think you were an iconoclast in college? Try being a tall, gay, banjo-playing fungus major with a microscope in your dorm room, walls decorated with fungal family trees you drew yourself.

That's who goes into mycology. Researchers in more glamorous fields, like, say, mammalian biology, have found almost all the charismatic megafauna out there and characterized it to within an inch of its life. The botanists have done the same for the plants. But the people who study creepy-crawlies are still slogging it out in the field. Beetles? Nobody knows how many more species of beetles are waiting to be discovered. And fungi? Oy. Depending on who you believe, there are between 1.5 million and 5 million species of fungi on earth. Only 100,000 of them have been named and characterized according to the (arcane, ancient) rules in the International Code of Botanical Nomenclature. Of those, barely a fifth have gene sequences in GenBank, the world's main storehouse of genomic data. Only a couple hundred have been sequenced completely, and most of those are yeast—because those are commercially useful.

Magnified fungi look like alien plants from a 1930s pulp sci-fi magazine cover, or a Dr. Seuss illustration rendered by Pixar. It's a weird landscape, not to everyone's taste. "If you found a new deer, you'd be on the cover of *Nature,*" says John Taylor, a mycologist at UC Berkeley. "If you find a new fungus, you're in the middle pages of *Mycotaxon.* But we're not bitter."

For hundreds of years, mycologists have named things the old-fashioned way — put a sample under a microscope and describe the shape of its body parts, the way it reproduced, the structure of its spores. The rules were typological, which is to say, a researcher had to have a physical specimen — called a "type" — stored in a herbarium somewhere, and sometimes an illustration of microscopic structure, too. And also a description in Latin. But that's all changing. The genomicists are taking over, and there's plenty of controversy over their methods — hoovering up thousands of genetic samples and sequencing them all at once. Scott, though, was trained by the old school of mycological taxonomy, the largely retired generation of guys who could ID a fungus on sight.

On the other hand, Scott didn't know much more about booze than he could learn from one of the bottles of whisky he kept in his office (which was also his bedroom). He had no idea that while winemakers have been storing their product in wooden barrels for centuries, the makers of distilled spirits didn't codify the practice of selling aged booze until the late 1700s. This new stage in the manufacturing cycle took the business of spirits to a new level. Now distillers needed real estate to warehouse the casks, and they needed a robust credit economy to fund the manufacture of a product that wouldn't be sold for years. A leisure class had to emerge that would pay a premium to drink something more refined than moonshine.

In other words, the birth of the economic ecosystem surrounding aged liquor represents a signal moment in the early Industrial Revolution, a mile marker on the road to a more civilized world. And somehow that fungus staining the walls of Lakeshore was a result of that journey.

• • •

In book I of *History,* Herodotus says that Armenian wine merchants in the fifth century BC, riding animal-skin boats with willow frames, could carry nearly 25 tons of wine along the Tigris and Euphrates Rivers to Babylon. Most wine histories say Herodotus wrote that the wine was transported in "barrels made of palm tree wood," the first example in history of storing wine in wooden barrels.

But Patrick McGovern — the archaeologist who identified the oldest evidence of human-controlled alcoholic fermentation — says that's a mistranslation. "If barrels were used by the Armenian merchants for their wine," McGovern writes, "they would have had to be made in Transcaucasia, and date palms do not grow there." The Armenians didn't have the wood to make the barrels. Herodotus uses the term *bikos phoinikeiou* to refer to the vessel holding the wine, and McGovern says *phoinikeiou* probably means "Phoenician," referring to a Phoenician-type amphora.

So who figured out that aging booze in wood makes it taste better? Today we know that barrels give aged beer, wine, and brown liquor — whisky, some kinds of rum, aged tequila, and so on — the distinct flavor of oak. It's a complex trait, and one that helps define those categories. But if McGovern is right, no one is exactly sure when people started doing it.

It wouldn't be crazy to pin it on the Romans. Their wine culture was evolved enough to have sommeliers — they called them *haustores* — and wine, *vinum,* classified into subcategories (sweet *dulce,* soft *molle,* white *album,* and dark red *sanguineum*). One of the whites, Falernian, was supposed to be aged at least a year before drinking, and older equaled better. At twenty to thirty years old it acquired a more amber color; in *Satyricon,* Petronius the Arbiter (somewhere around 27–66 AD) describes a dinner at which a honeyed wine is served first, followed by a 100-year-old Opimian Falernian. Booze chemistry changes over time even in the absence of wood, though, and evidence suggests the Romans aged their wine primarily in ceramic amphorae, sealed and heated — an early version of pasteurization, killing the microbes that spoil food without messing up the food itself.

Excavations of Roman storehouses called *horrea,* in Rome and the city of Ostia, suggest that the Romans built a specific kind of warehouse just to keep the copious amounts of wine they drank, as distinct from, say, grain or other dry goods. But the typical mode of storage was probably in a ceramic jar called a *dolium*—bigger and rounder than an amphora—buried up to its neck to keep it safe and its temperature constant.

The Romans had wooden barrels, though. Maybe they used them to store wine. Some historians argue they could have acquired the practice from the Gauls or Celts; McGovern thinks those people might have passed on the knowledge to the Minoans of Crete in the Bronze Age. Along with the signs of tree resins he found in a sample of wine from Turkey—probably there as a preservative, which is why the resins frankincense and myrrh were such valuable gifts from the Magi—one of McGovern's colleagues found a chemical called a lactone. Most likely, he says, it was b-methyl-g-octalactone, often found as a derivative of oak wood or resin. In the booze trade, it's known as "whisky lactone"; it's the chemical, derived from aging in wood, that gives red wine, Scotch, and cognac their coconutty, round, lush mouthfeel. Its concentration goes up if the wood is toasted—which is what coopers do to make barrels.

Did the Minoans add oak chips to their wine to flavor it, or age it in wood? No one has ever recovered a barrel from a Minoan site, but as McGovern points out, the Minoans knew how to build ships. And if you know how to bend wood into a curved shape that keeps water out, you have the skill set to bend wood into a curved shape that keeps liquid in.

Lactone is barely the beginning of what happens to alcoholic liquids stored in barrels. The chemical changes inside a cask—because of wood, air, and time—are every bit as profound as those that take place during fermentation and distillation.

Distilling really took off in Europe in the 1200s, and the Irish monks who first tried it probably kept their stock in wood barrels—which meant the booze soaked in the wood while it waited to be sold. Many histories of bourbon say that by the late 1780s US whisky was aged, taking as evidence that the barrels it was shipped in were marked "Old

Bourbon Whiskey." People liked the taste of that stuff, and figured it was because of time spent resting in the barrel. But "Old Bourbon" referred either to the counties in Kentucky where it came from or to the street in New Orleans on which it became popular—historians disagree on which—not to its manufacturing process. Anyway, all those early drinks were just as likely to be sold "green," or unaged. Routine aging in oak barrels, says bourbon historian Charles Cowdery, was still several decades away.

Brandy from the French region of Cognac typically spent a year or two in a barrel, and Americans in the late 1700s and early 1800s were Francophilic enough to associate this with quality. Cowdery's research says that some whisky in the United States was advertised as "old" in 1793, and by 1814 a few distillers were reporting the age in years of their product. But most whisky was still "common," clear white dog mixed with water to reduce the proof (if you were lucky—often it was adulterated with much worse stuff; food purity laws were still a way into the future, too).

People knew that you could set fire to the inside of previously used barrels to remove the taste of whatever they'd carried. But one of the rules for making bourbon is that the barrels have to be new, and they have to be charred. Cowdery doesn't know where that idea came from. "Charring used barrels before storing whiskey in them may have become so common, and the benefits so well known, that new barrels intended for whiskey storage were routinely charred too," he writes. "The fact that no record has been found to show that a radical change in the common practice occurred indicates that whatever changes took place probably were evolutionary."

By the 1840s, the evolution was complete. Bourbon, by law, had to spend time in a charred, new oak barrel. In Britain, similar rules were in effect for what people were allowed to describe as Scotch, though the barrels didn't have to be new. (Single malt makers tend to age their whisky in barrels previously used for sherry or bourbon, and sometimes "finish" it in casks used for more exotic stuff, like Madeira or rum—and you taste all of that in the glass.)

Wood barrels had shifted from being medium to message. Instead of being mere vessels for transporting beer, wine, and spirits, they became an indispensable part of the process of their manufacture. Something in the wood, combined with time, was the difference between a light, tart, mildly alcoholic grape juice and a mellow, rounded wine — between moonshine and bourbon, between tequila blanco and tequila añejo.

It's only in the last three decades or so that researchers started trying to figure out what, exactly, made the difference. The process of making a barrel changes the wood. The constituents of the wood leach into the liquid inside. The time that liquid spends in the barrel changes how it tastes when it comes out. But . . . how?

The only hints that something more than just office work is happening inside the beige building in an office park behind Santa Rosa's tiny airport are two big, metal chimneys. This is the Northern California headquarters of Tonnellerie Radoux, which started making casks for the French wine industry in 1947 and five decades later opened a satellite cooperage in NorCal wine country, an hour north of San Francisco.

Master cooper Francis Durand looks very, very French. When we meet, he's wearing a gray cap and a fleece-lined work vest over a black mock turtleneck. He also has a neatly trimmed, graying goatee. And an accent.

Durand takes me through a heavy door opposite the building's glass entrance, onto the factory floor. Beneath a thirty-foot-high ceiling, new wine barrels roll along rails getting sanded and finished. Durand leads me to the right, to the warehouse: 4,500 barrels stacked six high tower over us. The cold air smells of sweet, fresh sawdust.

In France, Durand was a sheep farmer. But his father made cognac, and Durand knew a little about woodworking, so in 1989 he left the farm to work for Radoux. In 1994, he came to California to start the US side of the business. "I was not speaking a word of English, and I had to work with people who know nothing about wood, machines, barrels, or wine." Even though he now runs a factory floor that produces 10,000 to 12,000 barrels a year — fifty-five barrels a day, if they're really cooking — he's still a craftsman. "To make a barrel, it's not very difficult, but it's so many steps," Durand says. "If you make one mistake, at the end, it's not a barrel. It's a box. If you're lucky."

The barrels are all oak, but that's nowhere near specific enough. At what Durand calls "the mother company," back in France, the coopers work with either fast-growing, thirsty Limousin or English oak, *Quercus robur,* or the more drought-tolerant, slower-growing *Q. petraea.* In North America, usually it's white oak — *Q. alba.* "It must be white oak, and a straight tree, long and high," Durand says. "Not those crooked California trees."

Radoux harvests from forests in Minnesota, Missouri, and West Virginia, in the Appalachians. That latter stand will be familiar to distilled spirits fans, since the rules say that if you want your mostly corn distillate to be called bourbon, you have to age it in new American white oak barrels. Most of those are sourced to Appalachia. Radoux only makes barrels for wine and high-end spirits, though Durand reserves a certain disdain for the way spirit makers think about wood. "I won't say a whisky barrel is a box, but it's not sophisticated like a barrel for wine," he says. "The wood is not aged. The toasting is not sophisticated."

Another difference: Bourbon makers, and therefore Scotch makers who use ex-bourbon casks, typically use barrels made from wood dried in a kiln to remove excess moisture before coopering. But wine barrels come from wood seasoned outside, exposed to extremes of temperature and natural fungi that infect the wood and begin the process of breaking it down and making it workable.

Wood is not the first material you might use to make a watertight container. Without sealants or adhesives — no tar, no varnish, no paraffin, no nails — it's even less appealing. From that perspective a barrel is an impressive bit of woodwork. It's a ring of wood staves, cut wider in the middle than at the ends and with a precisely calculated bevel along the long sides so that, when bent and encircled with metal bands, liquid can't get out — not through pores in the wood, not through the joins between the staves.

Structurally, wood is almost entirely composed of just three polymers — lignin, cellulose, and hemicellulose. Only a few organisms on earth break them down naturally, including the bacteria that live in termite guts and some species of fungus. None of the three are particularly soluble in ethanol or water. But somewhere between 5 and 12 percent of wood is other stuff. The hemicelluloses will degrade into simple sugars and then into furan compounds, and lignin eventually turns into phenols. Oak heartwood has eight different kinds of ellagitannins, the chemicals that contribute astringency to wine but also a sort of big and meaty flavor.

Those tannins leave visual evidence in the rings oak forms as it grows. Durand pulls a stave off a pile and shows the end to me, pointing at the striations with a ballpoint pen. "The white one is spring growth — soft wood, low tannins," he says. "And the dark one is summer growth, dense, with a lot of tannins." Unlike the structural polymers, oak tannins — different from the ones found in grapes — break down in water. "The trick is to select by the size of the growth ring," Durand says. Coopers used to do it by eye, intuiting which staves would be more tannic. Now, at Radoux, they use infrared spectroscopy.

Durand used to hand-cut and -fit the staves, but now he uses machines to assemble and bind them loosely with a temporary metal ring. The result looks like a giant wooden flower or one of those paper blossoms chefs put on the ends of lamb chops. Coopers heat the interior to about 200°F over an open flame while wetting the outside of the wood — nothing catches fire, but the heat changes the thermoplastic properties of the wood so it bends. The outer layers stretch but the inside doesn't compress. Once the staves cool off, they don't try to take their old shape — the bent wood stays bent, and stays tight.

Despite the hard labor required, the process is surprisingly precise. Three big digital clocks with red numbers note each barrel's time on the fire from high on the wall, and an instruction manual for the whole process is tacked above a workbench, each page in a clear plastic sleeve — written in English and Spanish. Lots of the coopers on Durand's line are Latino. "It's always the same guy who does the final step," Durand says. "He almost doesn't need the thermometers. He knows by touch."

Durand walks me over to a newly finished barrel, fresh off the fire, and invites me to stick my head in and take a whiff. It's amazing, like a bakery in a pine forest.

From there, the barrels get a groove called a croze cut in the top and bottom, and end-pieces placed inside, held by a paste of flour, water, and sawdust. "You can use wax," Durand says, "but flour is an old French tradition, and it's a very cheap product. It's natural, no flavor, and it reacts like wood."

Temporary, rough metal rings get replaced with shining ribbons of galvanized metal, cut and bent inward on a big machine and punched with holes for rivets. The barrels roll on rails under a belt sander and then are tested with pressurized water for leaks. By the time they're finished they look like something you might buy at Ikea—smooth, blond wood and bright, shiny metal, all with graceful curves. They look light, somehow, even though the biggest barrels weigh more than me. "No glue, no nails, nothing," says Durand, running a finger up a join on a finished barrel. It's a loving caress.

The Canadian mystery fungus had Scott stumped. He knew it was a heavy drinker, but he couldn't figure out how it was getting to the bar. In November of 2001 he got a break. Scott told his favorite wine importer, a sommelier-in-training, about the black-coated warehouse. The importer got it immediately. "That's the angels' share," he said.

Scott had never heard the term, but if he'd ever been on a formal distillery tour, he would have. Distilled spirits evaporate while they're aging—through the wood of the cask, or maybe the joins and bunghole. Whisky makers assume they'll lose 2 percent of volume a year, though it varies according to weather conditions and alcohol content. Poetically, they call that loss the angels' share, a portion of spirit offered up to heaven in thanks for a miracle. It's no small tithe—one particular fifty-year-old cask of Balvenie single malt lost 77 percent of its original volume by the time the distillery bottled it. (Of course, angels' share evaporation concentrates everything left inside. According to one taster's notes that Balvenie was "a magnificent dram" that built "slowly and majestically." I've tasted fifty-year-old Scotch once or twice, too, and while it seemed to have fewer layers of aroma and taste than something younger and more typical — a famously peaty whisky from Islay had almost no peat aroma left, for example — the flavor was indeed deep and focused.)

Slow-growth timber, with rings closer together, has higher levels of extractives available to the spirit. But whisky makers, at least, have never much cared about that, focusing instead on storage conditions — heat and humidity, primarily. At Jim Beam, for example, legendary distiller Booker Noe oriented his long, nine-story aging warehouses (called rickhouses in Kentucky) on a north-south axis, so the long sides would get sun for more of the day, and they'd heat up faster. Even atmospheric pressure can have an effect. "Flavors run along the cellulose chains from high pressure to low pressure," says Jim Swan, a famed barrel consultant. "The alcohol runs along the cellulose chains. Water can do the same. Some of the finest Scotch comes from places where there's a gain of water in the cask." Four years of aging on Scotland's west coast can actually result in a gain of five liters of water. In chilly wine caves and the low, single-story warehouses of cold, damp Scotland, barrels lose more ethanol; in the hotter American South, they tend to lose water — a factor exacerbated by the metal-sided rickhouses. They get much hotter on the high floors, which is why "small batch" or "single barrel" bourbon can be so interesting — those casks are selected for being accidentally better than the barrels around them.

Broadly, though: spirits aging in a warehouse give off ethanol vapor. Now Scott understood that the smell outside the warehouse represented a chemical gradient connecting the fungus to the distillery. That led him to his first suspect, the "cellar fungus" *Zasmidium*. It grows on the inside surfaces of wine-aging caves, accumulating in thatches that become home to other microbes. (At some wineries, oozing biofilms of bacteria drip down onto the barrels. It's sort of gross.) How many fungi could there be that lived on ethanol vapor? Scott figured his warehouse was harboring a giant *Zasmidium* colony. "Based on the similarity of the habitat, the little bit of physical descriptions I could get, I thought that's what it was," he says.

The Centraalbureau voor Schimmelcultures in Utrecht, Netherlands—the world's most important depot for fungal samples and genomes, more colloquially known as CBS—had cultures of *Zasmidium,* so Scott ordered some and put it under the microscope. It didn't look anything like the warehouse-staining fungus. Plus, *Zasmidium* only grows inside, in the cool, controlled climate of a cave. Scott's mystery fungus grew outside, over wide temperature ranges.

He was stumped. All he had was an educated guess that his mystery fungus was part of a group called the "sooty molds." One of the world's foremost experts in sooty molds, a researcher in his eighties named Stan Hughes, happened to work at Agriculture Canada in Ottawa, just down the hall from Canada's National Herbarium, one of the largest collections of fungus specimens in North America. Scott booked a flight to Ottawa.

Hughes was only too happy to receive him. Mostly retired now, Hughes still keeps an office on the second floor of a building that looks like a 1930s elementary school, set amid parkland and a few standard-issue squat government labs south of the government center in Ottawa. The office is oddly empty; Hughes has given away his books. Instead of a desk there's a workbench covered with copies of journal articles. A tall green metal cabinet in a corner is decorated with pictures of cats torn from calendars, and there's a microscope on a board propped up on twin piles of phone books. It seems to have been there a long time; Hughes, slightly hunched by age, has to stand on tiptoe to look through it. With his wisps of white hair and an illuminated magnifying glass hanging from a silver chain around his neck, Hughes looks every bit the Gandalf of fungi that Scott needed. He was happy to help with Scott's search. "It promotes the use of herbarium mycology," Hughes says, "as opposed to all the chemistry stuff." Geneticists need not apply.

Scott showed Hughes his samples from the distillery, and they rooted around for a couple of days in the herbarium, tall metal file cabinets packed together on rails, their shelves stacked with hand-labeled, hand-folded paper envelopes and matchboxes full of mold and mushrooms. Hughes remembered a sample someone had sent him, a piece of asbestos roofing tile from 1950s Denmark that was coated in black fungus — the same stuff, microscopically and to the naked eye, that Scott had seen in Lakeshore.

But they had a problem. The fungus was supposed to be something called *Torula compniacensis,* literally "Torula from Cognac." *Torula* was a junk genus, a drawer to throw fungi into when they didn't fit anywhere else. It makes mycologists shake their heads like plumbers frowning at a homeowner's attempt to patch a pipe.

Scott knew he'd have to do more than look at *Torula* with a microscope. He'd need to trace the literature back to the beginning, too. And what he found was confusing. In 1872, a pharmacist named Antonin Baudoin, director of the Cognac agricultural and industrial chemistry laboratory, published a pamphlet on a mold blackening the walls around distilleries in the region. Baudoin thought, incorrectly, that it was an unnamed member of the algal genus *Nostoc* and didn't try to give it a species name.

Then Charles Édouard Richon, a mycologist at the French Botanical Society, got wind of Baudoin's research. In an 1881 paper, Richon and a coauthor dinged Baudoin for serious mistakes and reclassified the fungus as *Torula compniacensis.* Richon gave some to a colleague, Casimir Roumeguère, who thought it looked like a fungus named earlier by the famous mycologist Pier Andrea Saccardo, but Saccardo had likewise gotten the name wrong, and Roumeguère further transcribed the name incorrectly in an influential exsiccata, a collectors' set of fungi samples that enthusiasts circulate to help stabilize nomenclature. Pretty soon a bunch of samples of the fungus from Cognac were floating around, all mislabeled.

Scott and Hughes traced the error back to its source. "And the herbarium in Ottawa happens to have some of Roumeguère's exsiccati," Scott says. "So Stan and I were able to go into the herbarium, pull it out, and see exactly what Baudoin had collected."

Under the microscope, what Richon called *Torula compniacensis* looked exactly like the samples from Lakeshore. But by a more precise modern definition, this stuff wasn't *Torula.* And more work in the herbarium showed that it wasn't like any other known genus, either. Scott realized he was going to get to name a new branch on the fungal family tree. But he had to follow the rules. "We needed a living culture that we could grow up," he says. They needed a new sample, an epitype, and it had to come from the same place as the original: France.

By coincidence, a colleague of Scott's, Richard Summerbell, was headed to a conference in Paris, and he was more than game for a paid trip to Cognac. He hit the jackpot at Rémy Martin. "When the tour was over, we got into the gift shop and bought our discount bottle of XO Rémy Martin, and an extra one for James, and we were released into the front yard," Summerbell says. "And there were some beautiful, black-covered bushes there. So we snapped off a bunch of dead twigs."

The discovery of a new fungal species might not make much noise, but a new genus — the next taxonomic category up the tree — is pretty cool. Scott and his colleagues nervously set about coming up with a brand-new name. He couldn't name it after himself; that's unspeakably crass. Hughes already had dozens of species and genera named after him (and anyway, naming a booze-swilling fungus after a teetotaler like Hughes didn't seem kosher). So the team settled on honoring the man who first brought the stuff to the attention of mycologists. They named the new genus *Baudoinia,* and they left the species name alone: *compniacensis.* In English: "Baudoin's fungus from Cognac."

Barrels full of booze are exciting places, chemically speaking. Of the structural components of wood, cellulose and hemicellulose are giant chains of repeating glucose molecules, and the heat of coopering breaks those into sugars — glucose, hexose, and pentose. But the third major component, lignin, is different. It's a massive molecule, too, but with nonrepeating subunits. About half of them are vanillin (vanilla flavored), and the rest is barbecue-flavored guaiacyl, clove-flavored eugenol, and syringaldehyde. At high heat, the spicy aromatic aldehydes in the lignin undergo Maillard reactions and yield the same flavors as browned meat. When it's hot outside, pores in the wood open up and the liquid moves inside, slurping up the tannins and other molecules that come from lignin decomposition. And the ethanol makes all those chemicals react with each other. The aldehydes mix with the acids and form fruity, tart esters.

Meanwhile, the evaporation of product out—the angels' share—means that external atmosphere gets drawn in. Sulfur compounds that didn't get picked off by the copper during distillation have a chance to evaporate or react into less offensive versions—though it can take years. Incoming oxygen means oxidation of ethanol to acetaldehyde and acetic acid. It's also what makes aging beer so tricky; beer is full of lipids, fat molecules, and those oxidize into a molecule called nonenal, which tastes like cardboard.

The liquid in the barrel even changes its essential molecular structure. Ethanol molecules cluster together when they come into contact with water. The number of clusters increases over time, and their presence can reduce the perception of ethanol in the final product. They can also stick to some volatile molecules, effectively making them less volatile and reducing their presence in the aroma of the spirit. When people call a drink "smooth," this might be what they're tasting.

Each oak species has different breakdown products. To figure out what works best, Jim Swan decided to try an experiment. Swan is something of a legend in the barrel world, an expert in wood who has consulted with, as far as I can tell, just about everyone. That includes Independent Stave, one of the main coopers for the US spirits industry. He got the distillers at Dry Fly to mature a batch of their white dog in each of Independent Stave's different woods—American, French, French-American hybrid, and European.

Swan presented some early results at the American Distillers Institute conference in Kentucky in 2012. At half a dozen top-ten round tables, audience members tasted about half an ounce of each differently matured whisky. The differences were subtle, but noticeable. The American oak–aged whiskies were more perfumed; the French oak had more vanilla and butterscotch. My favorite of the five (one was a "control," just a regularly produced bourbon cask) was the hybrid, with American oak staves and French oak heads. (Relative to the staves, the heads of a cask have a lot of surface area, which means more exposure of the liquid to whatever's in the wood.)

Swan thought the experiment needed more time to run. "French and European oak is more porous than American oak," he told the crowd. "It takes more time. And right now we're not looking at oxidation." You can't just bubble atmosphere or bottled oxygen through the barrels to try to accelerate the process, Swan says. That's O_2, two oxygen atoms stuck together, but if you want to oxidize aging spirit you need atomic oxygen, plain old O, a reactive atom looking for something to bond to.

A year later, I talk to Independent Stave's director of R&D, David Llodrá. Even more results had come in by then. In the American oak, vanillin levels had spiked and the so-called smoke phenols like guaiacol had gone way down across the board. To Llodrá, those results speak to the need for a kinetics of aging, a model that distillers can use to predict what'll happen over time as wine or spirits make their way deeper into the wood. "That's why a second-fill barrel tastes different," Llodrá says. "You've already washed off the surface, and with the second fill you go deeper into the wood. You're getting things out in a different ratio than you did in the first fill."

Can you get even crazier? The answer in boozemaking is always yes. If you don't care about getting the appellations of your region — if you don't mind not being able to call yourself "bourbon" or "Scotch whisky," then you can stop chopping down oak trees and look for something more exotic. A cooperage called Black Swan, based in Minnesota, makes barrels and liner inserts out of nontraditional woods, bored into a honeycomb pattern to increase surface area.

On the exhibition floor at the ADI meeting, the company had four small Mason jars at its booth. Each one contained white dog — at 59 percent ABV — and a piece of honeycombed wood about three inches long. They'd been steeping for a month, and all four were as brown as bourbon. Each had a different wood — white oak, cherry, yellow birch, and hickory. The birch had a weird roundness, but the standout was hickory, which had every barbecue flavor except smoke. I wasn't the only one to think they'd be good in beer. Black Swan, I learned, had just used hard maple inserts to age a 20,000-gallon batch for Ommegang, a New York brewer of Belgian-style ales.

Distillers and winemakers would love the aging process if it didn't take so long. Winemakers adore the flavor of oak in whites—sometimes overmuch—and the round mellowness of the best reds, but they can also sell young wines. Distillers, if they wanted to, could stick to unaged stuff like gins and eaux de vie and white rums, producing product as fast as they could sell it. But the brown liquors are, rightly or wrongly, seen as something of a pinnacle of distilling achievement.

Major distillers like Rémy Martin or Glenlivet or Jim Beam are hugely capitalized and, until the great whisky shortage of 2012 and 2013, they all had a deep back catalog. They used to be able to let what they make sit for three years, eight years, fifteen years, or more because they did the same thing three, eight, fifteen years ago and so have product to sell right now. The global popularity of whisky is squeezing those makers—more and more of them are switching to bottles with, as they say in the trade, no age statement. They don't say how old it is, because their marketing has equated age with quality, and what's in the bottle now includes spirit from younger barrels because they don't have anything else to sell. As rum researcher Rafael Arroyo wrote in 1945: "The ever increasing demands of the trade, the lack of adequate working capital, the anxiety for immediate returns, immoderate and unfair competition and many other influences of businesses compel the manufacturers to place their products on the market in the shortest possible time."

So every producer of booze, major and craft alike, has a common fantasy: make something that tastes old without aging.

As early as 1817, the industrial handbook *The Cabinet of Arts,* for example, recognizes in its chapter on brewing and distilling that time is the most important component of the flavor of fine French brandy. Its authors then go on to suggest that you should totally fake it. Distill a spirit as pure and flavor-free as possible, they say, and then add "oil of wine" — a wine concentrate — and some treacle or caramel for color. Presto! All the brandy, none of the wait. Golden-age bartender William Boothby recommended an equally dubious workaround in his 1891 *American Bar-Tender* for beer that seemed insufficiently aged: "Add a few handfuls of pickled cucumbers and Seville oranges, both chopped up," he writes in a chapter called "Valuable Secrets for Liquor Dealers." "This will make malt liquor appear six months older than it really is." Gosh! A whole six months, Mr. Boothby? Thanks! (The very same wording appears in a book called *Our Knowledge Box: Or, Old Secrets and New Discoveries* published sixteen years earlier, making Boothby seem even sketchier.)

In the late 1800s, whisky makers in Scotland began to adopt a practice borrowed from sherry makers in Spain, "seasoning" barrels with paxarette, a sweet, dark, sherry concentrate. They'd spray the stuff into the barrel under pressure to coat the inside, and then pour out the liquid. Cognac makers used to put their eau de vie in a sealed container and heat it, gradually, to between 140 and 175°F; the process was called *tranchage* and, while it accelerated oxidation without the concomitant angels' share evaporative losses, the eventual brandy tasted weird. The technique isn't legal anymore.

In his rum research, Arroyo collected a pages-long litany of tricks and techniques to artificially age rum. It included a few things you'd expect, like adding fruit extracts, sometimes alcoholic and sometimes merely aged — including "prune wines from Scotland." Oils of clove, cassia, vanilla, and bitter almond all made the cut, as did various forms of sugar, including maple, honey, and dextrose. Some rum makers got even more creative. They used heat, or alternated heat and cold. Or they bubbled oxidizing agents like oxygen, hydrogen peroxide, and ozone through the spirit. Arroyo says people get even more exotic, trying electrolysis and ultraviolet light.

Meanwhile lots of companies sell oak staves, chips, and filter-paper envelopes — oh, let's just call them tea bags — full of oak sawdust. Vintners or distillers dunk them in steel tanks to reproduce the chemistry of aging. In fact, in a paper on artificial aging technologies in rum since 1990, the alcoholic beverages group of the Department of Nutrition and Bromatology at Spain's University of Granada identified eighteen, and most involved the addition of oak chips.

That doesn't mean there are no more tricks to be had. A Taiwanese whisky called Kavalan — made at a distillery built with Jim Swan's help — has been winning competitions against classic single malts despite just two years of age on its spirit; its makers attribute its perceived maturity to high heat and humidity in Taiwan. It makes sense, theoretically — high heat drives the new-make spirit into the wood and accelerates all those other chemical reactions, and the humidity would preferentially pull ethanol out in the angels' share, maybe. Distilleries across the Southern Hemisphere, in India, Australia, and South Africa, are trying the same approach.

Because small distilleries without stock on hand want to turn out product faster, they'll often simply use smaller barrels, just 2 or 3 gallons compared to the 52 gallons in a typical whisky cask. The smaller barrels are more expensive, but the increased surface-to-volume ratio means faster extraction from the oak — three to five months instead of years and years. That's what Tuthilltown Spirits in New York does with its five products, sold in cute little 375-milliliter bottles for really quite a lot of money under the label Hudson Whiskies. When Ralph Erenzo and Brian Lee started distilling at Tuthilltown they planned to use little barrels just to get started, and then graduate up to more typical sizes as they grew. But "when we started using bigger barrels, like 5, 10, or 15 gallons — still real small — we started noticing that the flavor profiles were different," says Gable Erenzo, Ralph's son and one of the owners.

People trained to think of eighteen-year-old single malts as the definition of whisky tend not to like the resinous, piney notes that come from small-barrel aging. Younger whiskies get all the extraction of a long decade in oak, but little of the esterification or oxidation, and little of the reorganization of the liquid's molecular structure. They're immature, spiky instead of mellow. The booze is brown, but tastes green. Kris Berglund, who runs the only academic distilling program in the United States, at Michigan State University, has dedicated a chunk of his off-campus lab to small barrels and temperature control. "It turns out a lot of bad things happen when you use small barrels. Our conclusion is it's not such a great idea," Berglund says. "The haircut you take on the angels' share is enormous."

Now Tuthilltown tries for the best of both worlds. They produce enough to age some product in larger barrels, and then they vat together the small-barrel stuff with the big-barrel stuff. Whisky makers will argue this point—if they're trying to make something that tastes *just like* a classic from Scotland, small barrels won't do it. But if they're trying to make something that tastes different, interesting, maybe even good? Well, sure. Maybe. Why not?

Tuthilltown has one more move. Figuring that increased contact between liquid and wood was a good thing—and recalling that in Scotland, warehouse managers roll their barrels and move them from higher racks to lower ones to rotate among all the environmental conditions in the racks—Gable's father Ralph came up with the idea to play music with a lot of bass to the barrels, to stir up their contents. The distillers moved speakers into the warehouse, a 1788 gristmill, and at night cranked up A Tribe Called Quest and dubstep. "We started noticing really deep flavors and aromas," says Erenzo. "We call it our sonic barrel maturation process."

Eventually a sound engineer took the distillery tour, noticed the speakers, and asked what was going on. The idea captivated him—a few days later the engineer showed up with a laptop and a measuring tape, and started running calculations on the barrels, the warehouse, and the speakers. He reset the playlist; now each cask gets a specific low-frequency wavelength. "It's not as much fun as it used to be," Erenzo says, somewhat wistfully. "It doesn't sound like a rave every night. But each barrel gets its own tone." Does it make a difference? Erenzo says they've run the products of the different-sized barrels through their gas chromatograph to see what was what, but they've never compared sonicated barrels to those left in silent contemplation.

That's not even the craziest thing anyone has tried. Industry rumors say Diageo once tried wrapping casks in plastic to lock the angels' share in, so they'd have more volume at the end. The company has never released results, but supposedly they were not good. The bourbon distiller Buffalo Trace has a rickhouse dedicated to experimentation — they call it Warehouse X — and despite the fact that it only has room for about 150 barrels, it uses different chambers — including one open to the air — to vary aging conditions like humidity, natural light, and temperature. Or how about this one: Jefferson's Ocean bourbon ages for almost four years on a boat. On the water. The idea being, getting sloshed around by the waves exposes more of the liquid to the wood more rapidly — and your angels' share losses are replaced by briny, sea-scented air. It gets good reviews.

Even the most traditional approaches have their sleights of hand. In the hills above Santa Cruz, California, Dan Farber makes well-regarded brandy and apple brandy under the label Osocalis. Apart from using California wines to make his eaux de vie (and not being in the south of France), Farber follows the appellation rules for cognac and calvados — which include, even in the old country, a perfectly acceptable maneuver to add the flavor of age.

A geophysicist by training, Farber surprisingly prefers his nose and palate to analytical technology. "I decided I wasn't going to approach this from a scientific standpoint but more from an artistic one," he says. The most advanced machine in the distillery, a big shotcrete-and-straw-bale barn with a roof planted with local grasses, is a pot still of the Charentais style, short and almost spherical with a long, graceful neck. Distillers often mounted them on carriages and wheeled them up next to French orchards and vineyards so that farmers could process their crops. Farber knows the practice well; he trained in Calvados, Cognac, and Armagnac. "I went and visited small producers in the countryside, and I fell in love," Farber says. "In Cognac, most of them make their money selling eau de vie to the big houses, but they still have their own small chai with the best of the best."

So he has tried to build a version of that in a little vale, down a winding road barely visible from the main (only slightly less winding) road, across a wooden bridge, amid mud and freely ranging chickens. Farber uses 90-gallon French oak barrels, the same as they use in Cognac, because he says that's the optimal surface-to-volume ratio. He moves his product among barrels, some that might contribute flavors and some that won't, letting varietal eaux de vie take on new characteristics over at least four years, blending together, tasting, aging some more. The chai is dusty, spider-webbed, and banded with black fungus—probably James Scott's *Baudoinia*. "You really want to encourage those flora that give character to the chai," Farber says.

The thing about double-distilled cognac, and its slightly rougher, once-distilled cousin Armagnac, is that they're at the opposite end of the age spectrum from bourbon. Conventional wisdom is that American whisky generally doesn't benefit from more than a few years in the barrel; French brandies start getting really good only after three or four decades. "They go on to become independent things on their own," Farber says. "With these aged things, that's the beauty of it. They become themselves."

The French do, however, nudge that self-actualization along. On one of Farber's early trips to Cognac, he was at a friend's distillery and saw a vat of thick, brown liquid. "What's that?" he asked.

"It's my grandfather's *boise*," Farber's friend said. "It's ninety-five years old."

Boise is oak extract, essentially a strong tea made from oak shavings. It and caramel—burnt sugar syrup—are allowable additions to cognac. "The smell was amazing," Farber says. But the taste was deeply astringent. As additions to very old cognac, boise and syrup add what Farber calls "finesse," a big taste and a lingering finish. "If you go to an artisan, even the syrup is twenty-five years old," he says. "It's a much better thing than adding oak character in small barrels." All these things, says Farber, are notes for a smart blender—old syrup, relatively young brandy, boise, and perhaps decades of marrying in a cask on top of all that.

Osocalis's XO brandy—which really is spectacular—blends together nothing younger than twelve years old, and sells for more than $100 a bottle. It's worth it if you have that kind of recreational cash lying around, but makes for a tough business model for Farber. "We didn't release a bottle for thirteen years. We couldn't. There was great stuff already out there," Farber says. He's talking about the other West

Coast brandy and eau de vie makers — Germain-Robin, Clear Creek, and St. George. And now a new generation of small distillers is ironically making it even harder on the old guard. "Novelty today has such cachet in the market. If people can say, 'Hi, here's my three-year-old craft brandy for sixty dollars,' it really discourages the longer-term exercises."

Even large distillers would love to have a shortcut around those longer-term exercises. That's where a company called Terressentia comes in. Based in Charleston, South Carolina, Terressentia uses a proprietary, mostly secret purification technology that mimics aging in spirits in a fraction of the time. It sounds like snake oil, but it has adherents, and to my palate seems to make a difference. "We're licensed as a distillery, but we do not distill anything," says Earl Hewlette, the company's CEO. "'Purification' is not a word that our regulators let us use, so we talk about 'cleaning.' But basically that's what's happening."

According to O. Z. Tyler, one of the inventors of the technology, they add oak staves to increase exposure to those sought-after extractives, but then also bubble oxygen gas through the distillate to encourage oxidation. Ultrasonic waves remove unwanted congeners — somehow leaving the ones you do want. Tyler started out trying to find a way to artificially age whiskies, he says, but now they've expanded to other spirits, too, including tequila and gin.

"Industry has really applied only three different technologies to improving the flavor of ethanol. One is filtration, one is multiple distillation, and a third is barrel aging," says Hewlette. "What our technology does better than any of those is the removal of congeners. Our gas chromatographs show significantly greater reduction in the minor alcohols and free radicals than any of the other three methodologies can achieve. The flavors, whether you're talking about cactus for tequila or sugar cane for rum or grain for bourbon or Scotch, are more apparent." In eight hours, he says, new-make bourbon tastes like it spent six years in the rickhouse. All fully patented, of course. "We've already won more than ninety medals with the professional tasters of the world against the best quality things that are out there," Tyler says.

Hewlette offers to send me some samples, including a six-month-old bourbon that he said had beaten Woodford Reserve and Knob Creek, two giants in that world, in a blind tasting. I agree to receive his free booze, and a couple weeks later, a cardboard box arrives by FedEx, marked "fragile." Inside are seven tiny glass sample bottles, their caps sealed with tape: a gin, a tequila, a citrus vodka, a rum, a brandy, and two bourbons. I open each in turn—well, I ignore the citrus vodka, because come on.

The gin and tequila are completely ordinary. If you'd told me they were premium bottles, I would have believed it, but I don't taste anything spectacular. The straight bourbon is, as Hewlette promised, very good. The six-month-old was the equal of any small-distillery, small-barrel-aged bourbon I've had, which is to say, it had the over-oaked taste of 1990s California Chardonnay and a little bit of the green, short-chain tannin of any brown liquor that hasn't spent enough time oxidizing. But for something that hasn't been through a traditional aging process, that's actually kind of impressive. (Of course, I didn't taste a "before" sample.)

A Terressentia-like approach is not an option for someone like Farber at Osocalis. Partially that's because tradition is part of his marketing, just as it is with a major distiller. But at a deeper level, his French-countryside-artisanal aspirations are part of the fun. The part of aging when an interesting, floral eau de vie made of Colombard grapes goes into a barrel and spends eight years not tasting right, only to come back even more complex? That's just cool. "This is not an equilibrium phenomenon. Time comes into it. And there's kinetics," Farber says, still a scientist after all—but one of the spiritual ones who sees magic behind the molecules. "Time is a variable here. We're prepared to wait. It's just the natural game with brown spirits. If you don't have patience, this is not the right game for you."

In the end, here is what naming Scott's warehouse-staining fungus explained: nothing. Scott placed it more correctly into a wider taxonomy, but . . . so what? A genetic analysis showed that *Baudoinia* was only distantly related to cellar fungus, even if both are eating the angels' share. Scott is still trying to figure out how the fungus uses the angels' share — it seems to trigger *Baudoinia* to produce heat-shock proteins, protective against temperature extremes, which might explain how it can survive such a wide range of temperatures in habitats ranging from Cognac to Canada to Kentucky.

Giving the fungus a name didn't tell the folks at Hiram Walker how to get it off their neighbors' walls, either. Scott suggested using saltwater. And the distillery — now part of the transnational liquor conglomerate Pernod Ricard — eventually lost interest in further research. They set up a fund to pay for power washing, and that seemed to keep them on the good side of the Ministry of the Environment. The Scotch Whisky Association's official position is that more prosaic fungi are at fault, according to a spokesperson (who declined to make any data or researchers available to explain further). A group of residents in Kentucky who live near rickhouses, and whose homes are getting that characteristic coating of what looks like soot, are suing.

Which leaves unanswered a really important question. How does a fungus that's at least 135 million years old, older than *Homo sapiens,* find a near-perfect ecological niche amid aging distilled spirits, something people have been doing for only a couple centuries? "It's an urban extremophile," Scott says. Typically we don't think of cities as being particularly extreme environments, but extreme doesn't necessarily mean exotic. Few places on earth get as hot as a rooftop, or as dry as the corner of a room in a house. Fungi live in both. Presumably somewhere in the world, naturally occurring *Baudoinia* lives adjacent to naturally fermenting fruit. Or maybe it's everywhere, near dormant, a sluggish loser until it gets a whiff of ethanol. Evolution is full of stories of animals and plants fitting into hyper-specific niches, as if nature somehow got the specs in advance. Presumably *Baudoinia* — and other weird, technophilic fungi, like the one that grows in jet fuel tanks, or the one that grows in microchip etching solution — were bit players on pre-human earth. And then we came along and built them bespoke microparadises.

Today, the office of Sporometrics, in a former industrial neighborhood of Toronto given over to new media companies and architecture ateliers, still has ongoing *Baudoinia* experiments in the tidy, small laboratory in the back. Scott thinks its spread might have something to do with dew, because ethanol is 2,100 times better at dissolving in water than in air. Maybe opening and closing the warehouse at different times of day, when there's no dew for the angels' share to settle into, would limit the spread of the fungus. "I don't know," Scott says. "Who knows? We may have just discovered something."

Scott has other projects. His *Baudoinia* work helped get him tenure at the University of Toronto, where he still teaches. Sporometrics is profitable, and he consults for the provincial poison hotline, figuring out whether the mushrooms that are making someone sick warrant a stomach pump or a new liver. He's looking at how household microbes influence childhood asthma. And he's bored by all of it. He wants to figure out how *Baudoinia* works.

In fact, one snowy day Scott and I drove about 100 miles north of Toronto to Collingwood, on the southern tip of Lake Huron's Georgian Bay, to yet another distillery, chasing *Baudoinia.* On Google Earth, Scott had seen black stuff all over the walls of the home of Canadian Mist, and he wanted a sample.

Just as in Lakeshore, the air in Collingwood was redolent of the angels' share. The walls, street signs, and trees were coated in mold, up to an eighth of an inch in some places.

When I wrote about Scott and *Baudoinia* for *Wired,* I said that Scott snipped a fungus-coated branch off a bare tree and threw it into the back seat of his Toyota SUV. But the truth is, he handed me the pruner and kept the car running while I bounded through ankle-deep snow to snag the sample. We didn't have Canadian Mist's permission to be there, and Scott wanted to drive getaway. I put the branch in the back seat, wondering all the way back to Toronto if the fungus was getting into his upholstery.

Under the microscope at Sporometrics, the sample didn't look anything like *Baudoinia.* "No way," Scott says, looking at the flat-screen monitor hooked to the scope. "What's all this?" He points at tiny white spores dotting the brown/black mass of fungus. He can see *Baudoinia,* but it has company. “It’s got these round, rough things, and these smooth hyphae.” Scott sits cross-legged in his chair for a moment, chin resting in his hand. He looks . . . stumped. Then he straightens up. “No. That’s great. It makes it even cooler,” he says, a smile opening up. He’s going to spend tonight making up some agar to see what grows.

Six

SMELL AND TASTE

ABOUT TWO DECADES ago, a bunch of economists at Princeton formed a wine-tasting club. They made a few rules: They decided that they'd meet on the first Monday of every month, no matter what — and if one of them couldn't make it, that person had to find a replacement. Figuring that some of the wines they'd be tasting were too good to spit out, they engaged a car service, because even though they had eight people they also planned to try eight bottles per session, and "try" meant "drink."

But the most important rule was that the tastings would be blind. Nobody would know what was in the glasses until the end, so the fancy labels wouldn't influence their judgment.

The Princetonians liked red wines best, especially those from the French region of Bordeaux, but because the club members were all academics what they really liked was data. It wasn't enough to just talk about how the wines tasted. They wanted to see if they could articulate some kind of reproducible results that would separate the good wines from the bad. So one of the founders, Richard Quandt, wrote a computer program that would run the appropriate statistical tests on their ratings — agreement among tasters, correlations, and so on. Quandt is retired now, with several microeconomics textbooks on his CV (as well as a book on racetrack betting and another on how dogs think). He curses a lot, in a soft Russian accent. "Every time we had a wine tasting we'd run the program, have some comments, and post it on the website," Quandt says. "If anybody wants to test any hypothesis about how wines are ranked, we are now up to 1,030 wines, and every month we add eight more."

The influence of the software and the group — both dubbed Liquid Assets — grew. They formalized their research as the American Association of Wine Economics, and started to publish a journal. The entire endeavor has turned into a streamlined locomotive of skepticism about the vast, lucrative world of wine tasting and reviews. It's not a train you want to get in the way of.

"The professional wine tasters, Jancis Robinson and Robert Parker and so on, they really just taste and give their opinion," Quandt says. "Many of us are trained statisticians and do econometrics at the university level. Agreement on wines could happen by accident, and so it is essential to establish statistical significance. Otherwise it means diddlyshit. But hardly anybody in the business takes statistical significance seriously." What Quandt is talking about is finding a way to mathematically validate the scores that famous wine reviewers — or even experienced tasters like himself — give to wines. Which is to say, Quandt wants to connect preference to the qualities of the wine, to its indisputable chemistry. If you know what's in a wine, you ought to know whether it's good or bad — objectively.

But you can't. No one knows how to do it. Nobody can even identify, with certainty, all the ingredients, all the molecules, in a glass of wine (or beer or gin or whatever). Nobody understands, exactly, why booze tastes the way it does, and why people like it. And nobody understands, exactly, how human beings actually taste things.

Now, to be fair, Quandt doesn't really think anyone, no matter how expert, can produce an objective assessment of a wine's quality, origins, manufacture, or taste. Because that's what his data tells him. "We are eight guys who are pretty experienced and have been doing this together for twenty years, and yet when we do the statistical analysis of our tastings, the amount of disagreement is extraordinary," Quandt says. "Since people don't have identical utility functions, they evaluate different characteristics of the wine. As a result, a wine that tastes fantastic to one person will taste shitty to another one. In a sense, it's hopeless."

You might even think that over time the members of the Liquid Assets group would have learned from each other and learned more about wine. If that was true, their amount of intra-group agreement would have increased, even if they didn't agree with outsiders. But no dice. The numbers show they disagree with each other today as much as they did twenty years ago. In fact, they pride themselves on it.

The difference between Quandt's group and professional wine reviewers, then, is that the pros attempt to disguise their subjectivity with — well, put it this way: Quandt's paper on the subject for the *Journal of Wine Economics* is called "On Wine Bullshit":

> Since there are many wine writers, and there is a substantial overlap in the wines they write about (particularly Bordeaux wines), it is important that there be substantial agreement among them. And secondly, what they write must actually convey information; that is to say, it must be free of bullshit. Regrettably, wine evaluations fail on both counts.

In defense of wine reviewers — and reviewers of any booze — it's really, really hard to talk about how something smells or tastes. Discussions of flavor, that combination of taste and aroma, are subject to the frustrating limitations of analogy. We use descriptors that rely on "odor-object metaphor" — not what something smells *of,* but what something smells *like.* "The vocabulary of olfaction," write neurobiologist Donald Wilson and psychologist Richard Stevenson in *Learning to Smell,* "almost invariably ties the odor to its physical source." Benzaldehyde tastes *like* bitter almond and cherry. Cherry and bitter almond taste *like* themselves. Or benzaldehyde. That's excruciatingly non-helpful if you've never tasted either of those things or, more likely, if my perception of cherries isn't like yours. Because how could it be? Our noses are different. Our brains are different. Your cherry need not be my cherry.

So how do we talk about booze? How do we connect the subjective perception of what we are drinking with the objective knowledge of what's in it and how it's made? Perhaps not surprisingly, booze researchers are getting close to solving this problem. Research into the taste of alcohol promises to explain the taste of . . . well, everything, really. Booze, more than any other foodstuff, connects the quantifiable real world to the messy version of it we all create in our own brains.

Quandt thinks that pros like Parker — or your friend who always makes a big show over the wine list at a restaurant — are essentially making it all up. Or, like some storefront psychics, possibly they think they know what they're talking about, when in actuality they've merely intuited their way into a con. "I find it impossible to imagine that somebody can take a composite impression of a wine and decompose it into eight components and identify each one individually — a little bit of tobacco and a little of honey flavor and a little bit of citrus and a little bit of wet earth and a little bit of molding bushes," Quandt says. Way back in 1937, the humorist James Thurber stuck it to this kind of pompous, meaning-free connoisseurship in a *New Yorker* cartoon showing a wine taster saying of a glass, "It's a naive domestic Burgundy without any breeding, but I think you'll be amused by its presumption." A wonderful Roald Dahl short story, "Taste," from 1951, similarly depicts a hyper-verbal wine snob as a con artist and a fraud.

If you really want to blame someone for the wine world's excesses of verbal fluff, you might point a finger at the writer Pamela Vandyke Price. Her 1975 book *The Taste of Wine* popularized such descriptors as "forthcoming," "pinched," "sloppy," and "vivacious" as opposed to something more specific, or more chemical. Booze reviewers have been doing the same thing ever since.

Price was — perhaps unknowingly — trying to solve a real problem. Talking about taste and smell is not like talking about, say, colors. In 2012, researchers in Israel mixed together dozens of aromas to create a smell they called "olfactory white," an inoffensive mix that always smelled the same no matter which thirty ingredients they included. To them, their mixture was the equivalent of neutral white noise, or the color white—an equalized mix of every wavelength. But there isn't an aromatic spectrum in which "pure" aromas mix to create more complicated ones, the way different wavelengths of light present different colors to the eye. Most people agree that a specific range of wavelengths is "red," and we don't have to further describe that light as "stop sign–colored" or "blood-colored." Assuming you don't have some pathological colorblindness, even if your red doesn't look like my red—and how would we ever know?—you'd still be able to disarm a ticking time bomb if I told you to cut the red wire.

But smell and taste are different. For those senses, the hard philosophical and scientific work is trying to find words that we recognize in common and can share with each other, so that we all know what we're talking about.

The first step is to acknowledge just how limited our senses actually are. For example, it's possible to ask how many different aromas a person is able to perceive at once—like when taking a sip of wine. In 1998, a couple of Australian researchers, Andrew Livermore and David Laing, hypothesized that even if people could identify the smells of complicated things like coffee or kerosene within a second or two, those same people wouldn't be able to identify any of the hundreds of components of those same scents.

Livermore and Laing set up a device called an olfactometer that could puff vapor from anywhere from one to eight different samples, all carefully controlled for concentration by bubbling through nitrogen gas. (Their olfactometer was controlled by an Apple IIe, which I feel obligated to point out was, at the time of the research, already fifteen years old, a Model T of computers.) The specific smells they chose—smoky, strawberry, lavender, kerosene, rose, honey, cheese, and chocolate—were listed on a computer screen. The only thing their twenty-six volunteers had to do was sniff the air coming from an exhaust port and say which thing they were smelling. With between one and four components, the subjects confidently and quickly picked out the smells in the mix. But if the mixture had more components, they slowed way down . . . and their accuracy dropped to nothing. After four different smells, the human brain basically chunks them all together, and that gestalt aroma becomes the identifying smell for a given object.

Laing and other colleagues got a similar result a decade later, when they trained a bigger group of subjects to be able to perceive and identify three tastes — salty, sweet, and sour — and three aromas (cinnamon, grassy, and nail-polish remover). Once their volunteers could reliably identify all those smells and tastes, the researchers ran them past in combinations of increasing complexity, up to five and six at once. The subjects could pick out the tastes, but the aromas were totally lost on them.

It's possible that Quandt is even more right about the limits of human sensitivity than even he knows. A study titled, simply, "The Color of Odors," will destroy your faith in anybody's ability to taste anything. Here's how it worked: three French researchers started with two wines from Bordeaux, a white made with Sémillon and Sauvignon grapes and a red made with Cabernet Sauvignon and Merlot.

The researchers first had a group of subjects taste both the white wine and the red, under white light in clear glasses, and write down all the words they could think of to describe each one. In this test it didn't matter whether the tasters perceived the same things. Inter-rater reliability wasn't a factor here — the researchers didn't care if tasters agreed with each other about the wine color and taste, just that each taster would consistently call one "red" and one "white."

Then the researchers took an odorless, tasteless extract of the grape-skin pigment anthocyanin and dripped it into the white wine, turning it red. And they called the tasters back for a second go-around, asking them to compare the white wine and the colored wine — the same wine, in other words, with red food coloring. The result was a taste-test catastrophe. Almost to a person, the tasters chose to use the same words for the white wine from the initial tasting on the white wine in the second. And they used the same words for the red wine on the red-colored white wine. They simply could not tell the difference.

Color alone — not aroma, not flavor — told them what to expect, and that's exactly what they tasted.

But what about experts? At nice restaurants, a professional walks you through a wine list, telling you what each one tastes like and which ones might go best with what dish. Surely these sommeliers must have a more highly developed palate.

The four-level training to be a master sommelier, according to the Court of Master Sommeliers, culminates in a grueling, three-part final exam including a twenty-five-minute practical test in which the applicant must identify six wines correctly — grape variety, country, district, and vintage. Thousands of people take and pass the first two levels of the testing every year, but only a couple hundred try the fourth level — the one with the practical test — and only eight or ten pass. Today there are just over 200 of these "master sommeliers" in the world.

Tim Gaiser is one of them. He doesn't work in restaurants anymore; these days he's a consultant for wine buyers. I go to see Gaiser at his house in San Francisco's Sunset district because — I'm going to own up — while I have a pretty good palate for brown liquor, when it comes to wine I'm not much better than "yes, that's good" and "no, I don't like that one." I've never had the knack.

We take seats at his dining room table and Gaiser passes across a page of guidelines that the Court gives aspiring master sommeliers to use as a framework for tasting. Gaiser calls wine a "shared hallucination" — not far off from how the science fiction writer William Gibson described cyberspace, actually — but still says people can find commonalities among their perceptions. "We can say Cabernet from the Loire tends to have more red fruit than black. It's got red flowers, pronounced green herbal stem, green tobacco component, there's a chalk soil, it's very acidic, it's not really tannic, and it tends to be very dry," Gaiser says. "You know all those things, you've been told that, but when you have enough experience with it and you can remember it, that determines expertise."

Like all sommeliers, Gaiser has a preferred method for tasting a wine, which he's happy to show me. He heads into his kitchen and returns with a bottle of Cabernet Franc from the Loire. Cab Franc, he says, needs food to really come through in its glory, but Gaiser pours us each a glass, picks his up, and tilts his at a forty-five-degree angle about an inch from his nose. He opens his mouth slightly, and inhales. "Someone like me, when I put my nose in the glass, it all comes pretty quickly. What I'm doing is generating images." Gaiser takes a sip, spits, and then a grid opens up in front of his mind's eye, he says, like a big, light-up schedule hanging from the ceiling of a train station. It appears at the lower part of his sightline, and all those flavors and smells he named come through, with associated images. "It's a little stinky barnyard," Gaiser says. "Then there's, like, violets. Like red flowers or purple flowers. There is a dirty, musty, earthy element as well as an inorganic element, like chalk."

The same thing happens with the wine's structural qualities — acidity, tannins, and finish. Gaiser's scale for acidity is particularly visual. "Mine looks like a slide rule that's about from here to here," he says, karate chopping the air in front of him to demark an invisible line about four feet long, "with a red button on it. And I just wait for the button to move and stop." It's as if Gaiser has trained himself to be a very specific kind of synesthete, intentionally converting one kind of sensory input into another.

Here is what I think when I take a sip of the Cab Franc: Yeah, this is pretty good, I guess. I thought it was a little sour and thin, but I almost always think that about Loire wines. I like Italian wines from Umbria. Sue me.

Of course, Gaiser had already broken the key rule of Quandt's wine-tasting group at Princeton: He knew what the bottle was before he took a sip. Would he have been able to be as articulate — or to identify it at all — blind? "What would my hit rate be? Probably 70-plus percent," Gaiser says. "If they are really good examples and they are classic wines."

That seems like a pretty big if. The wine world is full of strange (and often delightful) labels and combinations. Gaiser admits that those could fool anyone, even a master. For him, the trick is finding ways not to eliminate subjectivity in tasting, but to share that subjectivity. "My strongest belief about wine is that it's not precise," Gaiser says. "We do everything we can to give structure to the experience."

More likely, he's filtering experience through memory and a trained vocabulary. In 2011 a team of researchers from the University of Padua in Italy and Macquarie University in Australia compared the discriminatory abilities of trained sommeliers to amateur wine drinkers and to sommeliers-in-training. It was a brutal test—the tasters were exposed to fifty aromas, ten of which were common household smells like shoe polish or garlic, and forty were wines. And of the wines, they were trying to specifically identify just ten Italians—five red and five white. The other thirty were there to distract their noses. Oh, and they weren't allowed to drink the wine—just sniff. Like I said: brutal.

The subjects got tested on their ability to accurately describe specific odorants—that is, use the right adjectives for them, as assessed by judges. And then they were asked to identify the wines.

As you might expect, the more training people had, the more descriptors they could call to mind. They had a bigger vocabulary to describe the aromas they smelled. But in terms of blind identification of what the researchers call "wine-relevant odorants," the sommeliers didn't do any better than the novices. The pros didn't have noses more sensitive or well trained than the amateurs. But also as you might expect, the pros and trainees were better at identifying specific wines, even against the background of the distraction wine samples. The amateurs knew they were drinking wine, but couldn't tell which one.

That actually jibes with Gaiser's description of his own experience. He and other sommeliers (and presumably the kind of people who can identify other exemplars of other drinks) are matching the new aroma or taste against a stored memory of what that class of drink tastes like. The difference isn't in an innate ability or skill at constructing and synthesizing "wine bullshit," in Quandt's construction, but in experience.

Some confirmatory evidence, albeit anecdotal: Years ago I went to dinner with a bunch of master distillers from big American whisky makers. The restaurant had a very competent single malt list, and so at dessert the pro I was sitting next to ordered five glasses of high-end Scotch. I'd tasted all of them before—maybe in younger, cheaper versions, but I was still confident that I'd be able to pick out which one was which based on smell alone. A few minutes later the waiter brought a tray to the table with five identical glasses on it, each filled with about three-quarters of an inch of golden liquid. Before the waiter could say anything, the distiller looked at the tray and named all five with perfect accuracy. Not by smell or by taste — but by color.

The really interesting thing about the Padua-Macquarie sommelier study, though, is that while the pros and trainees did better than the amateurs, they didn't do *a lot* better. The novices got an average of 7.5 out of 10 right, and the pros got 8.6 out of 10. "We tentatively suggest that the verbal skills, which are developed around wine, perhaps lead to a somewhat similar overestimation of confidence in expertise," the researchers write. They're hinting that knowing many words to describe wine makes people think they're better at identifying it than they really are. So go strong to the hoop with that wine list next time you're at a fancy restaurant, because the odds are the sommelier isn't a whole lot better at discriminating among wines than you are.

When you take a sip of wine, you're tasting a lot. The tongue is covered in taste cells — clustered into onion-shaped structures that we call taste buds. At the top of those cells are receptor molecules, chains of protein that sense external conditions. When the right molecule hits, the cell goes through all kinds of internal mechanics that lead up to giving adjacent nerve fibers a little squirt of chemicals called neurotransmitters, basically saying, "Hey, I got a taste here — let the brain know, wouldja?"

By the way, those taste buds aren't organized into any kind of map, the way you might have learned as a kid. Not only are the four basic tastes — sour, salty, sweet, and bitter — perceived all over the tongue, but some cells pick up a fifth taste, meaty umami. And some flavor scientists think more basic tastes remain to be characterized. One candidate: kokumi, or fattiness.

In a given glass of wine, anywhere up to about 15 percent of the molecules are going to be ethanol. (In a sip of cask-strength whisky ethanol might be more than half.) And ethanol is a quirky tastant. It activates receptors for both sweet and bitter, but also acts as an irritant, perceived by an entirely different mechanism. Receptors in the mouth called polymodal nociceptors—which means that they sense a lot of different things—pick up pain (or its less severe cousin, itch), extremes of temperature, and chemical irritants. These aren't taste buds; they work mostly via the trigeminal nerve that runs around the orbit of the eye, past the sinuses, and into the jaw and tongue. That's how we perceive the fire of chili peppers, usually through the chemical capsaicin, and the cold of mint, conveyed by menthol. (A little pain mixes into both of those as well.)

Ethanol is a teeny molecule, relatively speaking, and it's a bit lipophilic—that is, it'll bind to fat molecules. Since cell membranes are mostly fat, ethanol can pass right through them. A very, very strong acid might never set off pain receptors in the mouth because the acid molecules are too big and too charged; they can't get through the lipid membrane. But like the fatty acids that make cheeses taste sharp, ethanol slips right through.

Of course, ethanol isn't the only component of a drink. All those other molecules that yeast make (or don't touch), or that make it over the top of a still, are in there, too. How do we taste something super-complex like wine or rum? The answer is in the nose. Flavor is a combination of taste and smell.

You sniff a flower, or food cooking on a stove via "orthonasal olfaction," meaning the aroma goes into your nostrils and up your nose. But the action of chewing, swallowing, and breathing while you eat sends molecules up through the back of your throat and into the sinuses—that's "retronasal olfaction." Some researchers think that if you don't chew and swallow, you're limiting perception. That might deeply screw up tests where the subjects spit out the tastant—like, for example, wine tasting. The Princeton wine club had that one right, too.

In any case, once the volatilized, moisturized aromas waft into the sinuses, they encounter a dense, wrinkled, mucus-covered square inch of tissue, the nasal epithelium. Underneath the mucus are the ends of neurons, and built into the ends of these are receptor molecules to detect aroma. Nobody understands these receptors completely — like all receptors they're proteins, big molecules with topologies determined by the ways their subunits, amino acids, attract and repel one another. Olfactory receptors thread back and forth through the membrane of the cells seven times, but most of the specific amino acids that make up those proteins are unknown. The sevenfold membrane crossing wasn't described until 1991, by Columbia University scientists Linda Buck and Richard Axel. It won them the Nobel Prize.

Receptor neurons bundle together into cables called axons, feeding up through holes in a perforated bone just behind the eyeballs called the cribriform plate. (In a serious head injury, the skull can shift, and the lateral movement of the cribriform plate shears those axons like a knife through spaghetti. Snip! No more sense of smell.)

Once through the plate, the axons connect to two projections from the brain called the olfactory bulbs. There, in blobs of neurons called glomeruli, is where the bulk of the computation gets done. Mice, known for their acute sense of smell, have just about 1,800 glomeruli — but 1,000 genes that code for olfactory receptors. That's a lot of perceivable smells. Humans have a seemingly pathetic 370 genes for receptors, but we have 5,500 glomeruli per bulb. That's a lot of processing power. It must be doing something.

The part of the brain that integrates all this information, the olfactory cortex, also gets inputs from the limbic region and other areas that deal with emotion — the amygdala and hypothalamus, among others. Processing of smells in the brain, then, is tied to not only the chemical perception of a molecule but also how we feel about it, and how we feel in general.

Every other sense in the body is, in a way, indirect. In vision, light impinges on the retina, a sheet of cells at the back of the eye that makes pigments and connects to the optic nerve. In hearing, sound (which is really just waves of changing air pressure) pushes the eardrum in and out at particular frequencies, which translate via a series of tiny bones to nerves. Touch and taste are the same. Some cell, built to do the hard work of reception, gets between the stimulus and the nerves that lead to the brain for processing. Some physical effect — air pressure, reflected photons, whatever — gets between the stimulus and the perception. It's all a first-order derivative.

Not smell, though. When we smell something, we are smelling tiny pieces of that thing that have broken off, wafted through the air, and then touched actual neurons wired to actual pieces of brain. Olfaction is direct, with nothing between the thing we're smelling, the smell it has, and how we perceive that smell. It is our most intimate sense.

When you bring a glass of wine to your nose and mouth, you smell the air in the headspace, and volatilized wine molecules jump onto your olfactory epithelium. You take a sip, and all those polymodal nociceptors pick up texture and temperature. You get "mouthfeel," a subjective measure of viscosity and astringency due in large part to the presence of tannins. It's actually the sensation of proteins getting stripped out of your saliva, and it's received by the trigeminal nerve endings. Your taste buds pick up the sweetness and bitterness of the ethanol, plus the broad flavors of everything else.

You swallow, pumping a whole new set of volatile organic compounds into your epithelium. That's where you taste the oak of the barrel, and a whole bunch of other associative words you've learned to connect with wine flavors — blackberry, leather, butterscotch, grass, green apple. And the ethanol itself changes those flavors, alters your judgment on them. (Nobody's exactly sure whether ethanol makes a red wine taste better, but from experience I'll tell you that while non-alcoholic beer can be pretty good, "de-alcoholized" red wine tastes like existential death.)

The flavors and aromas of everything else in the glass kick in, too. In a mixed drink that might be a range of different distilled spirits, all with their own chemical signatures, plus fruit juices, plus sugar. Some molecules that wouldn't dissolve on their own, or in water, will be soluble in ethanol, making them more accessible to the taste cells and the olfactory epithelium as the warmth of the mouth volatilizes them.

And if you're drinking something carbonated? In 2009, a team of researchers at Columbia University, led by a taste expert named Charles Zuker, figured out that the same cells that pick up sour flavors on the tongue make enzymes that turn CO_2 into bicarbonate ions — as in baking soda — and protons, which are really just hydrogen atoms stripped of their single electron. Protons are one of the hallmarks of an acid; the tongue perceives them as sour. So what's really going on is the conversion of CO_2 into carbonic acid, in a backward sort of way. That's why flat soda pop tastes overly sweet. The sour CO_2 isn't there to balance it out.

Making all that even more complicated where smell and taste are concerned, ethanol has what researchers call "postingestive effects," which is to say, it makes you feel funny. Animals tend to learn to associate those effects with the taste. The flavor might be repellent, but the effects are attractive — so we start to think of the flavor differently. Or, conversely, maybe you're one of the human beings who doesn't make enough of the enzyme alcohol dehydrogenase in your liver — like 30 to 50 percent of Asians — so just a little bit of alcohol makes you as sick as someone who drinks a lot. Or maybe you had a bad experience with cheap rum in college; vomiting is what the taste researcher Alexander Bachmanov describes as "a particular case of Pavlovian conditioning."

Oh, and, of course, ethanol is addictive, so you could get to a place where you didn't like the taste, didn't like its effect, and didn't want to keep drinking it — but still felt compelled to. "For most animals, sensory perception of ethanol is aversive," Bachmanov says. "But if they are exposed to the consequences of ethanol, it's rewarding. And once animals are exposed enough to rewards, there's a conditioned flavor preference."

Bachmanov, a slim man with a beard and a gentle Russian accent, is a researcher at the Monell Chemical Senses Center in Philadelphia, one of the world's preeminent institutions for studying the science of smell and taste. He studies how rats and mice acquire a taste for ethanol, a chemical that wild-type mice and rats won't have anything to do with. Bachmanov's theory — and this is controversial — is that *no one* actually likes the taste of ethanol. "Well, I think so," he says, biting a nail. So why does anyone drink it? "I think the reason is that ethanol has pleasurable effects after it's consumed."

Bachmanov is right to be nervous. He's suggesting, essentially, that liquor stores, wine connoisseurship, home brewing, cocktail culture, the full contents of *Wine Spectator*, and everything that comes out of a sommelier's mouth other than spit is essentially designed to get us over the fact that alcohol tastes bad. This is more than Richard Quandt–scale bullshit. This is epic bullshit. Can Bachmanov prove it?

"If you don't mind being exposed to the smell of mice," he says.

We head over to the lab side of the building, and . . . wow. The air smells like fried ammonia, like a weaponized Chinese restaurant. It's normal; the mice are healthy and well kept. I try not to flinch.

Behind a locked door, inside a clean, dry room the size of a walk-in closet, are metal shelves full of shoebox-shaped plastic tubs. In the tubs: mice. Wire lids hold mouse kibble and two glass tubes — serological pipettes, to be technical. For this particular experiment, the tubes hold different concentrations of sugar solution given to different types of mice. If they drink more of tube A than of tube B, you know they prefer it, and the amount of the difference tells you by how much.

The two-tube test doesn't account for postingestive effects, though. To get around that, Bachmanov has a "lickometer." Actually, he prefers the less dirty-sounding name "Davis gustometer," but whatever you call it, it's a cage sitting next to an apparatus that can swing various bottles within the mouse's reach. Each bottle moves into position for a certain amount of time and then swaps out while an attached computer measures how many licks the mouse takes. "The whole test may take ten to fifteen minutes, and animals are exposed to multiple solutions," Bachmanov says. That's so fast that postingestive effects don't kick in. The mice either like and lick or don't and don't.

So, according to Bachmanov, humans are like mice. Most of us perceive four things in ethanol: sweet, bitter, a burning sensation, and a slightly unpleasant odor. At a relatively low ethanol concentration, like 10 percent — that'd be a serious beer or a relatively weak wine — most mice won't drink the stuff. The bitterness and the odor outweigh the sweetness.

But mice hyper-sensitive to sweetness act just the opposite. They love ethanol. And this isn't only true for mice. Human alcoholics report liking concentrated sucrose solution more than nonalcoholics, too.

Bring the ethanol concentration up to 30 or 40 percent — closer to that of a distilled spirit — and now all mice get repelled again. Even mice with that strong preference for sweet can't get over the burning or the bitterness. But right up to that point, if you retrain the mice to dislike the taste of sweet — by spiking a sweet drink with lithium chloride, which induces nausea — they become averse to alcohol, too, Bachmanov says.

He tried a bunch of other tests. Maybe the mice were craving the relatively high calories in ethanol. Maybe they were tasting some other quality. But none of those hypotheses panned out. The only predictor of whether the mice would like ethanol was whether they perceived sweetness more sensitively. So regardless of whether you describe those mice as being unable to taste bitterness as acutely or overtasting sweetness, the point is that only the weird ones liked the taste of alcohol.

As Bachmanov's fellow Monell researcher Bruce Bryant points out, at those higher concentrations ethanol is also an irritant. Thanks to polymodal nociceptors and ethanol's sneaky back door into cells, it actually *hurts* to drink, though you can get used to the sensation. Bachmanov's conclusion is that the postingestive effects outweigh those aversive tastes and sensations. In fact, super-tasters — who perceive every taste with greater intensity (sodium chloride tastes saltier, citric acid more sour) — perceive ethanol not as sweeter but more bitter. And super-tasters report lower levels of alcohol consumption than plain old tasters. That supports Bachmanov's claim, too — people whose sense of taste is more sensitive avoid ethanol.

Bachmanov's research actually backs into the history of cocktails. People have always added sugar, or sugary stuff, to make lousy booze taste better. In Bachmanov's construction, every drink other than vodka is really just a glass full of misdirection to get us around the fact that we hate the taste of the key ingredient. The congeners, the other tastes and smells, are delicious sauce on an unpleasant main course.

The thing about studying taste in rodents, though, is that all they can tell you is whether they like something or not. They can't describe what they're tasting. We all taste and smell food. But if we want to render more serious critical judgments about it, we have to be able to tell each other what the food tastes and smells like.

People have been trying to find a common language for tastes and smells, an aromatic equivalent to the electromagnetic spectrum, for centuries. In 1752, the Swedish botanist Carl von Linné—the taxonomist better known as Linnaeus—tried to structure a taxonomy of smell, connecting every aroma to a few basic scents. He failed, but other researchers followed, each with a new schema. In 1916, a German physiochemist asked volunteers to classify 415 different odors along the surface of a virtual prism with "fragrant," "putrid," and "ethereal" at the vertices of the triangle on one end and "spicy," "burned," and "resinous" at the other. It didn't work. A decade later, chemists Ernest Crocker and Lloyd Henderson tried to link what they thought were four basic smells to specific chemicals—"fragrant," for example, was benzyl acetate, and "burnt" was guaiacol. Their model predicted an upper limit of 10,000 perceivable aromas. It, too, was nonsense.

The booze scientists broke the logjam. Their conceptual breakthrough was an "aroma map" of all the things one might smell or taste in booze, organized into a pie chart: a wheel.

The pioneering geographer of alcohol's highways and byways was Morten Meilgaard. Born in Denmark in 1928, Meilgaard was an analytical chemist who specialized in yeast, and he became something of an evangelist, traveling all over the world studying beer.

In the early 1970s, Meilgaard set out to classify all those flavors. For all the components of beer he could parse, he calculated a ratio of how much of the compound was present to the level at which a trained drinker would perceive it. Meilgaard used those numbers to build an infographic, a multicolored wheel split into pie slices with all those flavors arrayed around the outside—green apple, mold, metallic, hoppy, and so on. It was a potent visual guide, so useful that in 1979 the European Brewery Convention, the American Society of Brewing

Chemists, and the Master Brewers Association of the Americas all adopted it.

Meilgaard's approach shot through the alcohol-making business. It had a special effect on another flavor researcher named Ann Noble, who in 1980 spent a sabbatical in England with a bunch of other flavor scientists — including Jim Swan, the specialist on whisky I mentioned in Chapter 5 — doing taste tourism. "We were all in a car going around to distilleries," Noble says. "Every time a smell would come around we were like dogs, like, boom!" Eventually Noble became a researcher at UC Davis; now retired, she still lives near the campus, in a sunny house decorated with musical instruments from various non-Western cultures. As we talk in her upstairs office, her half rottweiler, half German shepherd noses at me, as if bored by the academic chatter.

Three decades ago, Noble was studying wines from Bordeaux, trying to find descriptors she could correlate with their chemistry. She was teaching a wine-tasting class, and the students would go around the room and try to describe what they were tasting. "I wanted us to have words so we could go faster." She started making lists of descriptors and sending them to other wine scientists, about half of whom responded with their own ideas and corrections. That helped later, too, when Noble needed their buy-in and support for the system she was slowly building.

The "Noble Wine Wheel" first appeared in 1984, and its more formal 1990 version is now a standard reference, translated into multiple languages. "The wheel is just words," she says. "It's descriptive because it's hard for people to describe flavors. But I can reproducibly train you to recognize very specific aromas." This was the key realization: the way to connect subjective to objective is to teach people to use the same words for the same things — as if everyone got to vote on a specific wavelength that we'd all agree to call "red" from now on. The language changes from metaphor to definition.

In some circles, Noble's wheel is canon. "There were those who said, 'Ann Noble said these are the only words!'" she says, mocking the same didacticism that Thurber, Dahl, and Quandt did. "I didn't say these were the only words. I just said these were some words you can start with."

More wheels followed. The Scotch Whisky Research Institute developed one, as did the Macallan distillery. They're not easy. "It's one of the hardest things I've ever done," says Nancy Fraley, a distilling consultant who spent three years working on an aroma wheel for craft whiskies from small distilleries. "I'm not going to make millions off it. I envision that if I've done it right and this thing works" — she pats a printout of her wheel — "it'll be used as a tool." Small distillers, with their freedom and willingness to experiment, have begun to incorporate all kinds of weird grains into their grain bill. By fiat, Scotch whisky only uses barley. Bourbon only uses corn, barley, wheat, and rye. But these new kids are experimenting with blue corn, sorghum, spelt, millet, quinoa, and teff. "The Scotch whisky wheels only deal with barley malt," Fraley says. "And we have a lot of distillers that don't exactly know what they're doing yet."

Aroma wheels have become a key tool in getting those distillers up to speed. And other fields use them, too. Wheels exist for tequila, cognac, and gin — and for perfume, cheese, chocolate, coffee, and even body odor. From field to field, the wheels give the nose a common tongue.

Once you have a shared language, you can connect it to the chemicals in the booze. At least, that's the idea. Some of that work is taking place at the Scotch Whisky Research Institute, a low-profile place about ten minutes' drive from Edinburgh's airport, on the campus of Heriot-Watt University.

Drinks are full of congeners, the flavors, colors, aromas, even other alcohols like methanol and isopropanol. Whisky has over 150 of them, some perceivable at a parts-per-billion level. (Gin, by the way, usually shows only thirty to forty.) And then there are the unidentifiable compounds, the things that the fancy analytical devices at SWRI's lab can pick out but that don't have names yet, that haven't been characterized chemically.

Vodka is the poster child for the unknown frontiers of flavor science. By law, no matter the substrate used to make it — grains or grapes or the classic potato — vodka is supposed to be nothing but water and ethanol; distillers can vary the ratio, but that's it. In other words, the standard bottle of vodka is just 750 milliliters of H_2O and C_2H_6O. Yet die-hard vodka drinkers believe that the purest vodkas really do differ in flavor. On its face, that claim doesn't make sense. Absent any adulterations, like the glycerol some makers reportedly add to increase viscosity, any high-end vodka should taste like any other. One hypothesis for why they don't says that at concentrations above 40 percent alcohol — that's 80 proof — H_2O forms crystalline molecular cages called clathrates, trapping ethanol inside. The researchers suggest that the length and strength of the hydrogen bonds that loosely tie the water molecules into those cages give vodkas their different flavors. No one knows how; it's not like there are taste buds for hydrogen bond strength.

If you could nail down the precise molecules and concentrations in a good single malt Scotch — or indeed any other spirit — think of the advantages. For one thing, you could stop spending money trying to get the other molecules in there. Maybe you could use a cheaper wood than oak for the barrels, or a simpler process for malting the barley. Maybe you could use modern climatic systems to accelerate aging and get the same flavor, or better flavors. You'd also have a chemical signature for authenticity, which would let you chase down counterfeiters and distillers making product labeled as one thing but actually something else entirely — like, for example, a sizable proportion of the "whisky" in India.

That's why the whisky industry built SWRI, to deconstruct and analyze a 350-year-old craft process. The alcohol business has built these kind of labs before — in the 1930s and 1940s, the rum makers of Puerto Rico sponsored a lab run by Rafael Arroyo, and his research and patents still help guide the industry today. The same goes for the three buildings' worth of experimental winery and flavor chemistry labs at UC Davis. If you want to know whose priority that research is, just follow the money. In this case it leads to one of California's preeminent winemaking families. The buildings at UC Davis are called the Robert Mondavi Institute for Wine and Food Science.

Flavor chemists have been using gas chromatography to break complex mixtures into their chemical constituents since the 1950s. Different molecules move at different speeds past a detector, producing characteristic output "peaks." The catch is that the molecules all have to be volatile, capable of diffusing into a gas. (A related technology, high-performance liquid chromatography, can break out nonvolatiles.) In the mid-1980s, the technology gained enough resolution to be of use to modern flavor chemists, and in the 2000s they started to get better at linking those specific peaks, the molecules, to specific flavors. "That's what we do here: Use quantitative descriptive analysis," says Gordon Steele, SWRI's director of research. His team puts together groups of people, trains them to taste whisky, and then those panels figure out what names they're going to give the flavors they perceive. This is the state of the art in taste research.

Typically at a distillery, maintaining consistency is the job of the master blender, who is expected to understand that product so well that he can maintain its flavor even if it is a blend that ten years prior consisted of different "reciprocals," the single malt whiskies that went into it. At SWRI the subjective analysis, experience, and sense memories of a master blender are replaced by high-performance liquid chromatography and gas chromatography–mass spectroscopy.

"Ideally what you want is a compound you can identify that has an aroma attached," says Steele. "The more mystifying ones, there's an aroma but not a compound." The human nose, it turns out, is more sensitive than mass spectroscopy—we pick up things the machine cannot.

After we look at labs dedicated to the chemistry of distilling, the aging process, and evaluating barley strains, Steele walks me into one that, at last, smells like whisky. There are bottles everywhere. Some of them are counterfeits, most with the kind of not-quite-English syntax on their labels that suggests an Asian origin. "Awards Whiskey," for example, comes in a square, slope-shouldered bottle that looks an awful lot like Jack Daniel's, complete with black label. Steele cracks it open and we all take a whiff. It smells faintly of the cold-and-flu aisle at Walgreens. Maybe it's made of barley. Maybe not. "But how do you prove that in a court?" Steele asks. "It might even have Scotch whisky in it."

Laboratory manager Craig Owen takes the bottle. He's younger than Steele, wearing a fitted, striped shirt with French cuffs and a tie. Owen inhales deeply over the neck. "It's got a big lump of vanillin in it, that's what it's got," Owen says.

Next door, the lab benches are filled with row after row of microwave oven–sized machines. This is where the courtroom-worthy analysis will take place. Steele and Owen walk me to one particular set of machines set up to run in sequence. There's a mass spectrometer, and then a gas chromatograph — nothing unusual there. But the output of the gas chromatograph, in addition to a computer screen to display spikes of different molecules, has what looks like a hospital oxygen mask sticking out of the end.

The name of the process is gas chromatograph olfactometry, and it was a critical advance in connecting organic chemistry to taste. The mask is a "nose-port"; someone sits here, nose to high-tech grindstone, and gets a whiff of every molecule as it passes by the detector. That person then writes down a description of what he or she is smelling using an "aroma palette" of those agreed-upon words — things like peaty, feinty, oily, floral, estery, and sulfury. Everyone smelling the molecules is trained so that if they get a hit of, say, strawberry, they click on "estery." Those get matched to the molecular peak. If you know what the molecule is, now you know what it smells like. Of course, it's still possible to get molecules with no smell and smells with no molecule.

The mask also comes with a humidifier. "It can get quite tiring on the nose," Owen explains.

So now we have a shared language to deal with the subjective experience of drinking, and analytical technology to break down the objective components of the drink. How do we connect the language that the brain uses to think about booze with the data that comes off of a gas chromatograph olfactometer?

The answer is math.

"Fifteen years ago I would have said to you that all sulfur compounds in wine are negative," says Hildegarde Heymann, a sensory scientist at UC Davis. In wine, sulfur shows up chemically as thiol groups and mercaptans like hydrogen sulfide. "But in the last fifteen years we found the volatile thiols that are the reason New Zealand Sauvignon Blanc smells like New Zealand Sauvignon Blanc. The gas chromatography work finally got to the point where they could see them. You could smell them — cat pee, passion fruit, tropical, grapefruit — they would have used those words, but we couldn't have attached compounds to them. And once we know what the compounds are, we can say, how can we go back and manage these compounds?"

When Ann Noble retired, Heymann got her job. Now Heymann is working on a synthesis of the vocabulary-word-based ideas of the aroma wheel makers with gas chromatography and a healthy dollop of statistics. Her first step is training a panel of tasters. She gets a group of a dozen people, sits them around a table in a conference room — or in the tasting theater in the building where she works — and lets them taste twenty wines. "We try to pick the two most difficult wines in that set to try first, so they feel like they're getting somewhere," Heymann says.

Then she asks them to describe what they're tasting. The rule is, tasters can't use words for which Heymann can't create a reference sample. So, for example, "delicious" isn't allowed, but "apple" or "violet" (the flower, not the color) is. Coming up with those words can be tough. "If somebody's never done it before and they come up with three or four words, they're doing great," Heymann says. "If they've done it before, it's twelve to fifteen."

Heymann writes all the descriptors on the whiteboard in the room and looks for overlap. Then the group does it again for two more wines. And so on, working to come up with common references "so that your violets and my violets are the same thing," Heymann says. In essence, she's building a new aroma wheel for the specific perceptions of the tasting group.

Then Heymann makes reference samples. "You come back the next day with every apple thing so you can find out what they meant," she says. If the panel came up with "tropical," she'll bring canned tropical fruit juice. "Apple" might be slices of red Delicious apples soaked in neutral alcohol and then decanted. Heymann lets the panel smell the references and then rate them, 1 to 9, on whether they match the aromas they smelled in the wines. The process is iterative; the panel keeps smelling wine and smelling references until they all agree on matches. It's an hour a day, three times a week.

(Who has time for such stuff? California state law says the panelists have to be over twenty-one, so few undergraduates can scam their way into free drinks. "We suddenly seem to have people of a certain age," Heymann says. A group of sixty-somethings from the same swim club has been signing up a lot, which suits her fine. "They show up when they say they're going to show up. They're interested.")

Eventually the panel has tried all the wines and agreed on all the references, which have replicable standards. "Then you tell them, 'OK, we're going to start.'" The panelists head upstairs to small booths at one end of Heymann's lab, where they taste wines and tell her which flavors they perceive—and Heymann then links those flavors to chemical analysis.

To sum up the point of all that prep: the panel has inter-rater agreement on aromas, and Heymann knows what they mean when they use specific aroma words. The odor-object metaphors are generally agreed upon and connected to actual chemical compounds. "We're calibrating the group to speak the same language, and we're calibrating the language to a standard that we can use as a translation device," Heymann says. It's consistency of meaning, and when meaning is consistent, you can do statistics with it.

An example: Heymann is interested in what winemakers call terroir, the traits of a wine that come from the place where it was made. It's a fraught term, because it seems to want to mean something about the actual dirt in the vineyard as opposed to, say, microclimate or local microbe populations. Heymann prefers "regionality," or maybe "site specificity." So she asked winemakers from ten different areas of Australia, Geographical Indications recognized by the World Trade

Organization, to send exemplar bottles of 2009 Cabernet Sauvignon —samples that they felt were the most typical representation of what they made. (It's a good gig when you can request wine for science.)

Then she ran the wines past a trained panel of eighteen people. And sure enough, after some complicated statistics, not only did their reference words match to specific compounds in the wines, but Cabernets from different regions actually had different compositions to match. Heymann has done much the same thing for Malbec grapes from California, Washington, and Argentina—this time starting only with juice and making all the wine either at UC Davis or a site in Argentina, to minimize variations in the fermentation. "The regions had characteristics on their own," she says.

Heymann links the aromas her tasters perceive to chemical analyses of the wines, using statistical models to figure out how close the tasters are getting to quantifiable differences. All that information gets mapped on a grid representing axes of difference for the wines —chemistry and perception.

The best part of this approach is that it cuts through Quandtish bullshit. Heymann is familiar with the work of the Liquid Assets group —perhaps overmuch. "Why don't they talk to a sensory scientist?" Heymann says. "They assume away all the complications." They can't get a statistical bead on their wines because they're evaluating wine for quality, and quality, she says, is a problematic concept. Heymann is trying to quantify perceivable aromas, pleasant or not.

Her methodologies are new. Across the booze industry, few makers of beer and wine and spirits widely employ rigorous, lab-based testing of their products as a way of improving them. Most do basic quality control and rely on the trained noses, tongues, and brains of a few longtime employees for the important parts. At the point where that chemistry and biology actually touches a consumer, things like the design of a label, the shape of a bottle, or the décor of a bar may actually be just as important as what's actually being poured. There's nothing wrong with subjectivity—when it comes to choice of drink, picking based on the aesthetics of wood paneling versus hyper-modern LED backlighting is perhaps no more or less valid than basing a purchase on International Bitterness Units or peat parts-per-million. People who teach wine-tasting classes often tell funny stories about how their students, even with training, prefer box wine in a blind test. And research shows that people say they enjoy a wine more if they know it's more expensive. Sure, that bottle of red from the little village you found when you and your first love got lost in Tuscany on that rainy night was the best bottle of wine the world has ever made. Just don't try the same bottle again alone, sitting in front of a *Star Trek* rerun.

Once, at a high-end Chicago restaurant that has since closed, a friend and I ate dinner ten feet away from a nationally famous food critic. He wrote lovely, personal essays about food and cooking for glossy national magazines, and sometimes went on television—the kind of person, in other words, for whom chefs show off.

We were so transfixed by the critic's table that I have no recollection of what we ate. Judging by the number of plates he got, the chef was adding tidbits here and there, putting his kitchen through its paces. The waitstaff piled Roman amounts of food onto that table. And with every turn of the wheel—old plates out, new course in—another bottle of wine showed up. I didn't see any of the labels, but each bottle came with bigger glasses than the previous. By the time the critic's table had reached the main course, the wineglasses were—I'm not exaggerating—eighteen inches high, and the size of goldfish bowls.

I wouldn't begrudge any restaurant or its patrons their theater. Given a wine order of sufficient gravitas, the wine stewards at one of my favorite places to eat in New York will decant the bottle at a big table in the center of the room and indulge in a showy "seasoning" of the decanter, taking a whiff and sip to make sure it's exquisite. The whole thing is about as discreet as a stage whisper, and everyone loves it—especially the person who ordered the wine.

But I also know that, objectively, theater is all that it is. A wine will change its flavor over the time it sits in a glass, sure, and decanting wine is a very good way of removing any bits of yeast or lees remaining in the bottle, but a zillion-dollar, flat-bottom crystal decanter doesn't aerate the wine any better than Tupperware would. Fancy glasses might make drinking wine more fun, but glass shape has a limited effect on people's perception of its flavor. "There is no science," says Maximilian Riedel, CEO of the company famous for its pricey, beautiful, grape- and spirit-specific glassware. "We're not talking about headspace or molecular flow. It is all about the winemakers and their senses." The people who make the products ask for a new glass shape, and Riedel asks them what shape they think will show off their favorite qualities. Today, in addition to glasses specific to grape varietals, sake, and different spirits, Riedel also makes a special glass for water. "Due to different rim diameters, the composition of the glass gives you various tastes, and when I say 'tastes,' also satisfaction. Some glasses, they dry out your mouth and you want to drink more. Others, they cool your taste buds," Riedel says. "You will be amazed that water can feel different on your palate."

He's not wrong. I would be amazed. But I'll give him this: people drink beer 60 percent faster out of pint glasses with curved sides.

To me, my dinner across the room from the famous critic illustrates the core problem with studying booze. The vast mystery of how to make a great bottle of wine or a delicious cocktail that appeals not just to the maker but the audience, and do it again and again, propels the business and the art. Add the emotional context to the postingestive effects of alcohol, the changes it actually makes in the brain itself — the subject of the next chapter — and booze becomes an even more special case. Our taste in drinks may have little or nothing to do with how we taste drinks.

On the way out of Heymann's office, my head full of techniques for tasting wine, I notice a small sign pinned to her bulletin board. In neat, sans serif type is an epigram from the British statistician George E. P. Box: "Statisticians, like artists, have the bad habit of falling in love with their models."

Seven

BODY AND BRAIN

According to his obituaries, the psychologist Alan Marlatt was a generous mentor, a controversial researcher, and a serial marryer. When he died in 2011, Marlatt was best known for advocating treatments for addiction that didn't require complete abstinence.

That's a poor summary of a four-decade career, especially because Marlatt also solved two of the most difficult problems in the study of alcohol — how to trick someone into not knowing whether they were drinking or not, and how to study their behavior in the aftermath. In 1973, Marlatt and a couple of colleagues published a paper entitled "Loss of Control Drinking in Alcoholics: An Experimental Analogue." They were trying to understand why some people are unable to stop drinking after just one. Is it the physiological effects of alcohol itself, or a learned behavior, no different than social drinking except in amount?

To make the experiment work, though, Marlatt needed a reliable placebo. In most scientific studies of the effects of a chemical upon a person, subjects anticipate receiving an active agent. But basic practice requires assigning people to either an experimental group or a control group where subjects experience every condition of the test except the specific thing being studied. It's how scientists isolate the specific thing they're testing from coincidence, time of day, or some other unknown effect. The control group gets everything but the kick . . . and Marlatt had no way of isolating the effects of alcohol on people, because everyone could always tell if their drink had ethanol in it. It tasted different.

Just floating a little bit of something like gin on top of a nonalcoholic beverage didn't work. The team decided to switch to vodka, because combined with a mixer, it's nearly tasteless. "We spent many a pleasant evening experimenting with different alcoholic beverages," Marlatt wrote a decade later, "until we came up with a combination that would work for our study." The perfect research cocktail turned out to be one part vodka and five parts tonic water, served cold. In tests, tasters couldn't tell whether their drink had alcohol better than chance. This turned out to be profound, because it meant that instead of having only two groups, drug (vodka tonic) and placebo (tonic), Marlatt could have four: expects-ethanol/gets-ethanol and expects-placebo/gets-placebo, as traditional in drinking experiments, but also expects-placebo/gets-ethanol and, perhaps most important, expects-ethanol/gets-placebo. Marlatt called it the balanced placebo design, and for the first time it meant that researchers could sneak people a drink or a not-drink, contra their expectations, and see what happened.

What they found shaped Marlatt's work for years to come. His team told their subjects that they were participating in a "taste study," asking them for subjective judgments of the flavors of drinks. Whether the subjects were occasional drinkers, heavy drinkers, or bona fide alcoholics (recruited by giving sign-up sheets to "cooperating hotel desk clerks and bartenders in areas known to be frequented by alcoholics"), none of them experienced loss of control unless they thought they were getting alcohol. Even the alcoholics drank no more than the non-drinkers if they were in the expect placebo/get ethanol group. Expectancies — people's perceptions of what would happen if they had a drink — were critical to the effects of ethanol.

At about the same time that paper came out, Marlatt moved from the University of Wisconsin to the University of Washington. In Seattle he founded the Addictive Behaviors Research Center, focusing on building even more realistic environments and contexts for his expectancy research. It wasn't just the smell and taste — or lack of it — that sparked people's behaviors regarding alcohol. It was socializing, sitting on a stool in a darkened room with music playing. You couldn't build a laboratory inside a bar, Marlatt realized, but you could build a bar inside a lab.

So on the second floor of an academic building at the university, Marlatt installed a long, narrow countertop and put glassware and bottles behind it. He lowered the lights, put in a stereo system, and added five stools along the length of the counter. And he built in two-way mirrors, cameras, and microphones. Marlatt called it the Behavioral Alcohol Research Laboratory — the BAR Lab.

It was a virtual environment to simulate drinking perception and behavior — a holodeck for getting buzzed. "The idea is that many different factors go into intoxication beyond the alcohol itself. Alcohol is a socially imbibed drug. Environment is a factor," says Kim Fromme, a psychologist at the University of Texas and a student of Marlatt's. "Having a drink alone in your dining room is very different from having a drink in a bar out with friends. Alan came up with the idea of emulating a bar setting — dark rooms, neon lights, music, everything but the cigarette smoke."

The Lab let Marlatt figure out a bunch of things about how alcohol affects people. He found that given enough cues, people in the expect-ethanol/get-placebo group would show signs of intoxication, like slurred words, facial flushing, and an increased probability of flirting with that attractive person over by the buffet table. At low-to-moderate doses, like what you'd get with social drinking, instructions designed to create expectancies are more effective when the subjects are distracted — by a party or an erotic movie (booze research is fun, right?) — than when the subjects have the chance to sit quietly and evaluate their own internal state of intoxication. In short, our state of mind affects what alcohol does to us, just as alcohol affects our state of mind.

Outright drunkenness is relatively easy to characterize, even though its symptoms vary from person to person, ranging from volubility to introspection, excitement to sadness. Concentration and coordination go out the window; maybe you slur your words. You get tired. You try to put your elbow on the bar and miss. That's what happens when you have around 17 millimoles of ethanol per liter of blood, or 80 milligrams per every deciliter — sometimes also written "0.08 mg%," or 0.08 blood alcohol content, the legal standard for intoxication in most parts of the United States.

At higher concentrations, ethanol is a classic depressant of the central nervous system. Between 250 and 300 mg/dl, for example, it's an anesthetic. You're conked out, insensitive to pain. At 400 mg/dl ethanol is a solvent; that level of ingestion is fatal.

Marlatt had figured out a system to do what researchers had been trying for decades — learn what happens *on the way* to 0.08. Everyone agrees that alcohol can have dangerous effects, that at high-enough levels it makes driving a car or operating machinery potentially fatal. Consumed often enough in quantity, it is addictive, and can cause incalculable damage to the lives of addicts and the people around them. Ethanol can damage the liver, the pancreas, the kidneys, the circulatory system, the brain. It's a psychoactive drug, with all the trouble that can come with that designation.

But what about consumption in lesser quantities? That might be harmless — even beneficial, either because a chemical that lowers stress levels a bit might actually be a good thing, or because a compound called resveratrol, mostly found in wine, in some studies curtails one of the primary markers for physiological aging. This isn't the realm of the binge drinker or the alcohol-dependent subject. It is, instead, a peculiar time centered around, let's say, the first sip of the second drink — the warm, spreading tingle, that slight sense that your brain is still looking at something even after your eyes have moved away. Maybe you're more confident, happier. You were tense; now you're relaxed. Your friends get better looking. Another drink seems like an even better idea. Or, for a not insignificant number of people, your stomach gets upset. You feel queasy. Your face flushes red, and maybe you get a little jolt of nonspecific anxiety. It's unpleasant. A second drink seems like a chore.

That level — what the researchers who study it call the "relevant range" — is in blood-alcohol terms between around 0.04 and 0.05. In social-drinking terminology this is the place between I'll-have-another-beer-please and no-thanks-just-the-bill. It's where the vast majority of drinking takes place, and a minimum of the research. And that's a shame. Marlatt made possible a whole new way of looking at the relevant range, because he understood that it was fascinating.

After 10 million years of ethanol consumption, 10,000 years of dedicated production, and over a century of focused scientific research, human beings still aren't totally sure what relatively small amounts of ethanol do to the body. The tiny molecule passes through cell walls like a ghost, making its way to almost every organ in the body. It's a pain-causing irritant (but also has numbing effects), and it's a decent source of calories (but not nutrition). It crosses the vaunted blood-brain barrier with ease and acts as both a stimulant and as a central nervous system depressant. Its effects vary within a single individual depending on the circumstances of its consumption, between individuals depending on genetics and experience, and between groups of individuals depending on genetics, environment, and tradition.

All after just a couple of drinks.

The volunteer identified only as *Ius* might have been the most patient human being ever to walk into a lab. He was the only subject in a set of experiments in the 1920s recorded under the title "Effect of Dilute Alcohol Given by Rectal Injection During Sleep."

Ius was about twenty-five years old, a student at Harvard Medical School, and a little guy: five-foot-six, 127 pounds. He was healthy, "used to taking small amounts of alcohol in the form of beer," and "very intelligent and cooperative." For six nights, at about 6 P.M., *Ius* came to the laboratory of T. M. Carpenter, of the Carnegie Nutrition Laboratory in Boston, and got an enema. Then the experiment started. Here's psychophysiologist Walter Miles describing the setup, because crazy always sounds better in clinic-speak:

> He lay down in the clinical respiration apparatus, which is of the closed-circuit type, with a chamber completely enclosing the subject while he is in a prone position. Pneumographs were arranged to register any body activity during the night, a stethoscope was secured in position over the apex beat of the heart for counting the heart-rate, electrodes . . . were attached to both wrists and the left ankle for the taking of standard electrocardiograms during the night, and the catheter through which the alcohol or control solution was to be passed was inserted in the subject's rectum. When the man had lain quietly for a considerable amount of time after the making of these adjustments, the cover of the chamber was lowered into position and the metabolism experiment began. These metabolism measurements were continuous throughout the night. By the use of a clock and an electric signal, a sound was periodically produced near the subject's head. If awake, he responded to this by pressing an electric button comfortably attached to the hand. Usually by 9 P.M. all was in readiness for the metabolism experiment and the chamber closed. The subject invariably went to sleep soon after that time. The rectal injection began about 2½ hours later. . . .
>
> Dr. Carpenter gave the injection of the alcohol solution or salt solution very slowly by the drop method, using a gravity fall of about 1 meter. The injection required approximately 2 hours and did not waken the subject.

To recap: professional researchers took a scrawny Harvard med student, taped primitive electrodes made of wet gauze to his hands and feet, stuffed him into an airtight coffin, played buzzing noises at him until he fell asleep, and then squirted booze through a rubber tube into his butt.

At about 6 A.M., Carpenter's team popped the top of the coffin, brought *Ius* out, and (after letting him pee and wash up) tested his metabolism and responses to various stimuli. Miles recorded his heart rate, how quickly he blinked after a loud noise, his ability to withstand mild electrical shocks to the fingers, and so on.

So what brilliant insight did *Ius*'s suffering yield? For what world-shattering data about alcohol did an Ivy Leaguer walk around funny for a week? The results, it must have crushed *Ius* to hear, were inconclusive. On the mornings where he got ethanol instead of saltwater, he was a little slower in general. Kind of intuitive, that finding.

The thing is, this field of research was so new in the 1920s that they weren't testing what ethanol did so much as checking to see if it did *anything.* Drinking to excess was prevalent on the job, but this was a postindustrial revolution era when "the job" often required either the precision and attention of a modern office or the handling of machinery with tremendous destructive potential. Planting corn while drunk is one thing; operating a steam shovel is another. Mostly Miles wanted to know if *Ius,* or someone drinking more normally, would simply be able to work the next day. One of the main proxy tests the Nutrition Lab did on buzzed subjects was to see how good they were at typing.

The researchers at the Nutrition Lab were trying to work out the basics of digestion and nutrition. They weren't even sure that people could absorb ethanol — or anything else — through the rectum. (You can.)

In fact, the rectum turned out to be a really good place to absorb ethanol, in the end. (Not sorry.) Drink something via the normal pathway — which is to say, through your mouth — and metabolism starts early, in the stomach and upper intestine. But run ethanol through the colon and it passes through the walls and into the bloodstream, allowing it to reach the brain more quickly.

Miles and his team didn't know any of this. That was the whole point. In 1913, the head of the Nutrition Laboratory, a researcher named Francis Benedict, had pitched his "Proposed Tentative Program for an Investigation of the Physiological Effects of Alcohol." Benedict soon turned a room at the lab into a fully kitted-out physiological research facility to look at the effects of ethanol.

Neuroscience was in its infancy; so, too, was medical imaging. Scientists understood some of the basic mechanisms of genetics but didn't know what DNA was, much less how to look at it. The testing apparatus was, necessarily, primitive. An L-shaped wooden device for measuring the patellar reflex was, for example, an assembly of magnets and pendulums that looks better suited to eliciting a confession than a knee jerk. When the eyelashes of the test subjects started getting in the way of recording their blinks and eye movements, Benedict and his colleague Raymond Dodge write, "It became necessary to standardize them by using artificial lashes. These were cut from black paper . . . Very heavy gum arabic solution proved a satisfactory adhesive medium." So, yes, they stuck paper eyelashes on people.

Benedict wanted a baseline measurement for human metabolism and response to all those weird tests so that he could contrast that data with what happened under the influence of ethanol. He wanted, in short, a reliable control on his experiments. It was the same problem Marlatt set out to solve sixty years later. When it comes to ethanol, whether you're studying human beings or rats or fruit flies, *everyone* knows what it feels like to drink alcohol. And that biases people's responses — what they say they feel, and what their physiology seems to do. Benedict's team tried a bunch of approaches to disguise drinks-versus-placebo — concentrated orange oil, chile peppers, sugar. Nothing worked. Encapsulating the ethanol would have required the subjects to swallow too many giant pills. Stomach tubes introduced their own biases, as did intravenous injection. Finally Benedict gave up; his subjects could almost always tell when they were getting booze. Unless, like *Ius,* they got it up the butt.

That was the kind of gadget-heavy experiment Walter Miles loved. The book he wrote to record his findings at the Carnegie Nutrition Lab is full of even crazier gear than Benedict's. My favorite is his approach to testing eye-hand coordination — he uses a "pursuit pendulum" that irregularly splashes liquid; the subject is supposed to catch as much as he can. (How did buzzed people do? Worse.)

After 275 pages of this kind of thing, Miles confirms the status quo. "A decrease in organic efficiency due to depressant action . . . the whole qualitative picture is one of decreased human efficiency as a quickly-following result from the ingestion of this pharmacodynamic substance," Miles writes. That's all we get. Alcohol slows you down, folks. Take it easy a little bit, maybe. It would be up to researchers down the line to work out what was actually going on.

From the first sip of an adult beverage, your body goes to work. The ethanol has to be oxidized, broken up and converted to a more usable form. As long as it's in your bloodstream as ethanol, you'll feel its effects; how long it takes to process depends on a lot of factors. Ethanol gets absorbed straight from the stomach and upper intestine, so if there's food in there, too, absorption goes more slowly. Drink faster and absorption goes faster . . . to a point. Higher concentrations of alcohol work their depressant magic on the GI tract, slowing your physiology down and delaying absorption. At those levels ethanol is also an irritant, causing your stomach to secrete mucus, slowing things down even more.

From the stomach, ethanol gets pulled mostly into the portal vein, a direct line to the liver. There, an enzyme called alcohol dehydrogenase oxidizes the ethanol, converting it to acetaldehyde. And acetaldehyde is a bastard.

Imagine a molecule roughly the shape of the Mercedes logo, an upside-down Y. At the center is a carbon atom, double-bonded to an oxygen atom at the top and with one hydrogen atom stuck to each leg. That's the preservative formaldehyde. Now, replace one of the hydrogens with other atoms or molecules, and you get other aldehydes. Stick a methyl group on there in place of hydrogen atom and you get acetaldehyde.

It's no problem in small amounts. But acetaldehyde is highly reactive—it wants to stick to other molecules. Those compound molecules, called adducts, gum up the works of pretty much anything they touch. Stuck to DNA, acetaldehyde forms at least one cancer-causing chemical and interferes with a process called methylation, a key way the body controls what proteins get made from what genes at what time. Acetaldehyde sticks to the microtubules that form the skeletons of cells, the collagen that holds our connective tissue together, and the hemoglobin that carries oxygen in our blood. It even gloms on to the neurotransmitters serotonin and dopamine, which may have something to do with how alcohol causes addiction, screwing with habit formation and our perception of pleasure.

The liver cells working on ethanol have to pull more than their fair share of oxygen from the bloodstream to make the chemistry work — they're playing a complicated shell game with electrons, adding them to and taking them away from a long sequence of molecules. But that chain ends with some free hydrogen ions, protons, and a need for more oxygen to attach them to, making water. The end result is not enough oxygen to go around; cells at the exit of the liver start to suffocate and become more vulnerable to toxins and pathogens. (We do get something from the process — food scientists measure nutritional density in kilocalories per gram, which dieters shorthand to just "calories." Your basic carbohydrate — bread, let's say — clocks in at 4.1 kcal/g, but ethanol nearly doubles that. Of course, those calories are largely empty, without vitamins, minerals, or proteins along for the ride. That's a good argument to drink beer, I guess. It's full of protein. Or you could order cocktails made with fresh juice, especially since people who drink get up to 10 percent of their total calorie intake from ethanol. Alcoholics get up to 50 percent.)

Of course, the whole point of the liver is to deal with toxins. "When you cut yourself, there's a response. Immune cells come in, and you get a little scar that forms and eventually heals. In the liver something very similar happens in response to injury. Immune cells come in to clear out debris and whatever was damaged, and you get a fibrotic response," says Laura Nagy, a pathobiologist at the Cleveland Clinic who studies how ethanol affects the liver. "When you get to higher levels of consumption, the wound-healing response can't keep up and damaged tissue remains." Ordinarily an inflammatory response like this one results from the immune system fighting off an infection, but an inflamed, ethanol-soaked liver is actually *more* vulnerable to disease. Nobody knows why.

Chronic alcohol use — and not even at binge levels — throws off one of the other key functions of the liver, the breakdown and metabolism of fats and fatty acids. They start to accumulate in the organ; "fatty liver" is a sign of chronic, heavy alcohol use and in extreme cases a precursor to cirrhosis.

Enjoying your drink so far? Hang on; we're getting to the fun stuff. First we have to get rid of that acetaldehyde, though. The liver makes a few different versions of an enzyme called aldehyde dehydrogenase — ALDH1 and ALDH2 metabolize acetaldehyde. How much of these enzymes your body makes, and how well, is one of the main determinants of whether or not people drink. About half of people of Han Chinese, Taiwanese, and Japanese descent make a version of ALDH2 that doesn't work at all. That's the basis of the so-called Asian flush response, the characteristic reddening of the face that some Asians get when they drink. It comes with some nasty tummy symptoms, and worse — Japanese drinkers in one study ended up with much higher rates of esophageal cancer.

In fact, the side effects of acetaldehyde accumulation are so bad that they're the basis for the first drug ever developed to treat alcoholism. Disulfiram — better known as Antabuse — inhibits aldehyde dehydrogenase. You can drink, and you can get drunk, but you also throw up. It's a potent negative reinforcer.

Ethanol that the liver misses gets back into the bloodstream. Within twenty minutes of the first drink, ethanol can make you have to pee, because it suppresses a chemical called vasopressin, a neurotransmitter that in the kidneys goes by another alias: antidiuretic hormone, or ADH. Normally it tells the kidneys to hold on to water in the body; without it, the walls of the tubules that make up the kidneys' infrastructure switch from sponge-like to something more like pipe. All that liquid heads to the bladder, and you pee it out, which means the body's overall concentration of electrolytes — potassium, sodium, and chloride — gets higher. In chronic heavy drinkers and alcoholics this does all kinds of damage and makes cirrhosis of the liver worse, but people who drink moderately actually see lots of kidney benefits.

The body has all sorts of physiological machinery, then, to deal with ethanol, as long as it stays below a certain rate of flow. Where things really get interesting is in the brain — ethanol does strange stuff up there. To begin to understand that, though, we have to go to a party.

Alan Gevins claims that he didn't set out to say anything about alcohol. As head of the San Francisco Brain Research Institute, Gevins is an expert in electroencephalograms — EEGs — the measurement of electrical activity in the brain. What he was really trying to do was test a piece of headgear that measures EEG portably and comfortably, and, most importantly, wirelessly. That's when the booze came in. "I needed to make the leap away from stereotypic tasks," Gevins says. "We're social creatures." He needed a way to record what's going on in the brain during real-world social activity — no MRI tubes, no brain scans. "So I thought, let's leap way ahead: ten people getting drunk."

EEG has a long history in the study of alcohol. Because it samples the brain's electrical activity quickly and repeatedly, over the course of milliseconds, EEG is a particularly good method for measuring changes over time (as opposed to the spatial results you get from imaging technologies like magnetic resonance). Some of the best work in the field came out of the lab of Henri Begleiter, who helped figure out that ethanol induced a greater response in what's called the "slow alpha" frequency band.

Gevins came up with a stretchy nylon cap attached to a bunch of electrodes. People wearing them look goofy, but not cyborg-goofy. "Alcohol is unique among different drugs we've tested," he says. "It lights up the brain like a light bulb." And even though ethanol is technically a depressant, as far as an EEG is concerned it's a stimulant — in the relevant range, walking right up to intoxication. The peak effect doesn't kick in until about an hour after taking a drink, but when it does, it's pronounced. Every area under the electrodes essentially synchronizes.

He gathered fifteen friends and colleagues — scientists, engineers, research assistants, administrators — and told them the deal. Sushi, canapés, Martinis, Sea Breezes, and the EEG hat. "I pride myself on my Martinis," Gevins says. "Olives. Very dry. I prefer Stoli." After some typical partying, he passed around Breathalyzers and made sure everyone was brushing up against 0.07 BAC or so (though a few folks, he says, got considerably drunker—beyond 0.10). And it worked. Gevins's hat, and the equations he came up with for reading the output, could tell if people were intoxicated.

It was a triumph for EEG and the effects of alcohol, and it bodes well for Gevins's wireless EEG hat, which he has now tested on north of 1,500 people. But while EEG research does a good job of showing the ultimate effects of ethanol on the brain, it's of little use on the proximate causes.

Other drugs are so much easier. Seriously, alcohol researchers pine for their straightforwardness. Go to work on the effects of heroin, an opioid derivative, and you know you're working on receptors in the brain built to go off when they sense an opioid—the brain makes its own versions, called enkephalins, as part of the circuitry that makes us feel good. (The opioid-squirting neurons themselves eventually project to the neurons that secrete dopamine, the neurotransmitter that our brains use to signal a reward, a good thing.) Heroin, morphine, and opium lock right into that machinery. Or marijuana? No problem. The brain has endogenous cannabinoid receptors, too.

But not alcohol. "The truth is, we don't know at the molecular level what alcohol is binding to. It's never been resolved," says George Koob, director of the National Institute on Alcohol Abuse and Alcoholism, behavioral psychologist, and one of the world's foremost experts on addiction. "It's just a small molecule that hangs out in the water part of our bodies and zips through the nervous system. That's why there are these changes as you get more and more intoxicated. It recruits more and more neurons, starting in the cortex and then moving in a wave down toward the reptile brain, where all the reward transmitters are."

In Koob's construction, the ethanol starts out in the frontal cortex, the brain's outer surface, and then moves to the hippocampus, where alcohol seems to preferentially affect neurons used for laying down memories. So when those are blocked, you get blackouts and memory loss. As the ethanol migrates even deeper, to the cerebellum, that causes motor coordination problems—stumbling, slurring words, that kind of thing. Right? "Actually, it doesn't make sense at all," Koob says. Because that story doesn't explain *how* ethanol moves through those regions, and what it actually does when it's there.

Possibly the problem is so challenging because nobody is really sure how the brain works in the first place. Neuroscience dogma is that the brain is a network of neurons — 100 billion of them, with 100 trillion connections, at best (or maybe least worst) estimate. Somehow that network does the computations that become our minds. Communications essentially take two forms: Do More (or, in neuro parlance, an excitatory impulse) and Do Less (called an inhibitory signal). An individual nerve can get hundreds, even thousands, of these excitatory and inhibitory inputs at the same time. Whether that nerve fires its own signal is the result of a constant summing-up of those inputs. So, do all that trillions of times and a brain becomes a mind.

Neurons don't touch; they stretch right up to each other and stop, the little teases. The spaces between them are called synapses, gaps just about a millionth of an inch wide — 20 to 40 nanometers. To tell each other what to do across the synaptic gap — to convey excitatory or inhibitory signals — neurons squirt chemicals at each other called neurotransmitters. Broadly, excitatory signals are carried by the neurotransmitter glutamate (same stuff as in monosodium glutamate, MSG, which puts the umami taste in umami). And the inhibitory neurotransmitter is gamma-aminobutyric acid, or GABA.

In the relevant range, ethanol blocks some glutamate receptors and activates some GABA receptors. It's a double-negative: turn down the thing that tells the brain to go, and turn up the thing that makes it stop. Now, GABA-squirting neurons often inhibit neurons that make those good-vibe-signaling enkephalins I mentioned earlier. So GABA, simplistically, keeps us from feeling too good all the time. Tune it down and you feel better.

Ethanol to opioid to dopamine to feeling good — "and there you have a pretty little circuit," says Jennifer Mitchell, a neuroscientist who studies addiction at the Ernest Gallo Clinic and Research Center in Emeryville. "So you say, OK, on some level opioids are involved in the effects of alcohol. That's pretty straightforward."

Mitchell is a fast-talking, brightly intense woman with dark hair and (on the day we meet for breakfast) three earrings in one ear, two in the other, multiple rings, and an orange scarf. She got interested in studying drugs in college at Reed, in Oregon, where an anarchic lack of a drug policy meant her fellow students used, and got screwed up by, a vast pharmacopoeia. Mitchell wanted to study suffering, and she saw a good model for that in people who went beyond experimentation and became dependent. She wanted to find out why.

Eventually, that research turned to finding a drug that would fight alcohol addiction. A couple already exist — one, Antabuse, is the aldehyde dehydrogenase blocker that makes you vomit. Another, naltrexone, blocks the brain's reward pathways, so a drink doesn't make you feel good anymore. Unfortunately, when you're taking naltrexone, nothing else feels good, either.

Both drugs operate too broadly. Mitchell, like a lot of other researchers, wants to develop a more fine-grained description of what ethanol actually does in the brain — and then block that mechanism, and only that. "When I was in grad school, the rat went into the box and I would stare at it and it would stare at me and I would make some comments and observations and put the rat back," Mitchell says. "You look at thirty years' worth of data and it's mostly in rats or mice, with an occasional ferret or guinea pig or hamster. Some cats. And you think, 'I have no idea what this means for the human condition.'"

Mitchell knew that she needed to look at some human brains — not addicts, not binge drinkers, but just people in the relevant range. "Moderately intoxicated, but not ripped," as she puts it.

No one had ever tried it before — because it's very, very hard. First, Mitchell had to figure out how to target a specific receptor for alcohol — and remember, no one is really sure what receptor that is. But people have theories. Like: the opioid receptor. It actually comes in three types, designated by the Greek letters mu, kappa, and delta, and the mu opioid receptor seems particularly important in recreational drugs. Breed a mouse without it, and it stops drinking alcohol or taking opioids.

Mitchell found a particularly potent opioid called carfentanil, thousands of times stronger than heroin. She didn't need it for its effects, though. Carfentanil is super-powered because of how strongly it binds to the mu opioid receptor, which means that if you track trace amounts of it in the brain, you can see if those receptors are being triggered. That tracking gets done the same way that Sebastian Meier tracked the molecular progress of fermentation for Carlsberg back in Chapter 3. You have to connect a radioactive atom to the molecule you're tracing.

So Mitchell found a National Laboratory with a particle accelerator that could make a radioactive isotope of carbon, C-11. She used that to tag carfentanil; now a positron emission tomography machine — a PET scanner — would be able to follow Mitchell's carfentanil as it moved through a person's brain, in real time.

The experiment, then, would go like this: Give people radioactive carfentanil and then give them a drink. If ethanol induces more activity in the mu receptors, they light up on the scan.

That sounds simple; getting permission to do it was not. Counting the National Lab, a couple of hospitals, and a couple of other locations, the experiment took place at half a dozen places, and because Mitchell was using human subjects she had to get permission from each of the facilities' Institutional Review Boards. After two years, Mitchell finally had enough subjects to make it worthwhile. In the end the size of her study population — N, as scientists say — was only 25 (13 heavy drinkers and 12 people who drank only rarely), all recruited via Craigslist.

Then a new question came up: how to get them up to 0.05 blood alcohol. The best way would have been to use a device called a clamp, which both injects ethanol intravenously and reads blood alcohol, titrating up or down as needed. But Mitchell couldn't get ahold of one, and anyway her subjects were already stuffed into a PET tube with an IV line injecting the short-lived, radioactive drug. Adding a clamp would have been impractical.

So Mitchell instead turned to a drink with a standardized amount of alcohol, calibrated to the subjects' weight and sex. (Women respond to ethanol almost twice as strongly as men, in general.) She bought lab-quality, high-proof ethanol and cut it with a little bit of juice, so she could administer as little liquid as possible. The subjects would be in the PET tube all day, and, as we've learned, alcohol suppresses vasopressin. "You don't think about it until the first subject is in the magnet and says they need to pee," Mitchell says. (Obviously none of Mitchell's subjects had the dedication of *Ius,* the victim of the rectal-administration-while-sleeping experiments.)

Once you have someone in the magnet, nicely buzzed, brains loaded with radioactive super-heroin, you can go hunting. But that's actually not quite true — if you do an imaging study without knowing what you're looking for, you end up with silly assertions about which parts of the brain do what, like make you a liberal or a conservative. To do worthwhile imaging research, even with an *N* of 25, you start with "regions of interest," parts of the brain you hypothesize are going to light up on the scanner. And you compare those to, in this case, control regions that shouldn't have anything to do with opioid receptors, and regions that have opioid receptors but that aren't involved in the reward system.

Mitchell's regions of interest did indeed light up. Most affected: the nucleus accumbens, which you'd reach if you drilled straight back from slightly below the center of your forehead, and the orbitofrontal cortex, which sits above your eyeballs.

"The orbitofrontal cortex was more of a surprise," Mitchell says. "It's more like cognitive control and executive control, executive function, making decisions, weighing two things to decide which you want. This is obviously very anthropomorphized, but the nucleus accumbens is much more like, 'I want, go get it, I'm motivated.' So you have this model where the OFC is telling the nucleus accumbens when to wait."

The hypothesis, then — still remaining to be tested — is that ethanol induces the release of endogenous opioids in these regions, providing a jolt of good feeling but also undoing a person's self-control. That leads to more drinking and worse decisions, which is why the relevant range is relevant. It's the beginning of everything that sometimes happens with more alcohol.

• • •

The song of the zebra finch has long been a model for language acquisition and usage, because elements of the songs are passed from parents to children. Presented with ethanol, finches will indeed partake, and sometimes even get their blood alcohol level up to the equivalent of a human 0.08. And once there, their songs — according to research presented at a conference (and therefore not peer reviewed, so this isn't rock-solid) — get kind of muddy. The songs become disorganized, their elements less precise. Put simply, give lots of ethanol to zebra finches and, just like humans, they'll start to slur their words.

In people, ethanol's numbing effects go right to work on the tongue and throat, making it difficult to distinguish a *D* sound from a *T* sound or to vocalize "sss" instead of "sh." Linguists call these kinds of changes segmental effects, but ethanol also has supra-segmental effects, like a slowdown in the overall rate of speech and changes in pitch and tone. Supra-segmental effects seem to be related to muscle control. Dropping or inserting words inappropriately is probably more neurological.

In fact, the effects of ethanol on speech are so predictable that even a computer can detect them. In 2011, an international computer speech group issued a challenge to researchers: develop software that would allow a computer to determine if a person was drunk, listening to speech alone. The coders who took up the task made some vague hand waves about how a functioning system could replace Breathalyzers in remote testing, but really they were more interested in the speech recognition challenge.

The rules: success would be tied to detection using a really useful database called the German Alcohol Language Corpus, a set of recordings of a few dozen German speakers getting drunk, and then recorded again two weeks later, sober.

"The idea is that when one is drunk, one speaks with a different accent," says William Yang Wang, a computer science graduate student at Carnegie Mellon University who led one of the teams. "We are basically assuming that each phoneme you speak in a sober state is very different than the phoneme in a drunk state."

Wang actually wanted to study how computers could determine from speech a characteristic called "level of interest"—how *into* something the speaker is. You can imagine pollsters or salespeople might find that skill useful. But the challenges and approaches to both problems are, he says, the same. "Maybe you stop on some words for longer when you're drunk. When you're sober you might not stop at that word. And when you're drunk you may emphasize a specific word," Wang says. So his team looked at things like speaking rate and rate of vocal cord vibration, known as fundamental frequency (usually above 180 Hz for men and 255 Hz for women, but with alcohol it goes even higher).

Wang's software worked about 75 percent of the time—not quite good enough to win the challenge, certainly not good enough to replace highway patrol Breathalyzers. It got more accurate at higher blood alcohol levels, pushing 0.08 and above. Down in the relevant range, he says, it didn't do nearly as well. A human being still has some advantages when trying to assess someone's state of intoxication, Wang says. "We're actually processing multimodal signals, which is an advantage over computers," he tells me. "When I go to a bar, I can look at someone's head positions, their gestures, and I can listen to their speech and look at what kind of words they use." That's why a bartender knows to cut you off well before a Breathalyzer does.

We've been talking about the physical and mental effects of ethanol, but other substances in booze might have psychoactive effects too. When people claim that tequila makes them mean or red wine gives them a headache, they're blaming congeners, and it's certainly true that the amount and types of congeners vary from drink to drink. They might also affect the brain and body differently.

The most famous story of a bad-acting congener is probably absinthe. At the turn of the twentieth century the French drank 9,500,000 gallons of the stuff a year—that's 48 million bottles—and it had become synonymous with a certain Bohemian, artistic hedonism. Absinthe takes a base distillate, either something neutral like vodka or a flavored spirit like brandy, and adds herbs and botanicals as flavoring.

The main contributor is anise; in that respect it's like ouzo, raki, Sambuca, and other licorice-flavored regional firewaters. But also on the list is wormwood, which contains trace amounts of a hallucinogenic compound called thujone. In the early 1900s, thujone was blamed for "absinthism," a syndrome that turned otherwise normal people into murderous epileptic psychopaths. (As recently as 1989, one researcher suggested in *Scientific American* that absinthism contributed to Vincent van Gogh's suicide.) By 1915, the drink was illegal throughout much of Europe and North America.

That kind of challenge never sits well with cocktail historians. After all, without absinthe, you can't make a Corpse Reviver #2 (gin, Cointreau, Cocchi Americano, lemon, absinthe rinse — though some people prefer Lillet to the Cocchi Americano), or even a decent Sazerac (rye, Peychaud's Bitters, sugar cube, absinthe rinse — arguably the very first mixed drink). As for wormwood-free anisette replacements like Pernod, well, they're fine for poaching shrimp. So in the 1990s, a New Orleans booze aficionado named Ted Breaux got obsessed with absolving absinthe. He set out to determine whether there was enough thujone in pre-ban formulations to drive people crazy.

After a few years of collecting absinthe paraphernalia — spoons for suspending a sugar cube over a glass, ornate carafes with spigots to drip water on the sugar cube and melt it into the absinthe, and so on — Breaux finally found a pre-ban book with protocols for making absinthe itself. So he tried to produce some. It was terrible, but he didn't know why.

Eventually Breaux managed to procure a bottle of the real deal. Using a syringe, he withdrew a sip through the cork and tasted it. It was worlds more complex than what he'd made. His pre-ban absinthe "had a honeyed texture, distinct herbal and floral notes, and a gentle roundness uncharacteristic of such a strong liquor," Breaux told *Wired*. "Those protocols were crap."

Gathering a few more pre-ban bottles, Breaux cadged time on gas chromatographs and experimented with reverse-engineered recipes until he had one that was all but indistinguishable from the old stuff — and in the process found that pre-ban absinthe had thujone levels of just about five parts per million, nowhere near enough to have a hallucinogenic effect. He also found that the alcohol levels hovered around a sky-high 140 proof. That goes a long way toward explaining how habitual consumption could give someone seizures and homicidal tendencies.

(One cool thing about absinthe: it's translucent and sort of greenish, but when you add water it turns cloudy, almost milky. One of the major congeners in anisettes like absinthe, ouzo, and pastis is an oil called anethol. Straight out of the bottle, it's at a kind of equilibrium with ethanol and water. But drip more water in and the ethanol diffuses into the water and away from the anethol, which then forms into larger droplets. It's a spontaneously formed emulsion, in other words, that snaps the liquid from translucent to opaque.)

Even if thujone doesn't make absinthe a murder potion, other congeners do have perceivable effects. In an experiment in the late 1960s, researchers gave people vodka, bourbon, and a "super-bourbon" loaded with even more congeners (which the researchers don't identify). The subjects' blood alcohol levels were the same, and produced the same effects on an EEG — smaller alpha-wave amplitude and more slow-wave activity. That's just being tired, basically. But in the super-bourbon group, the effects lasted longer and were more pronounced.

So what causes that? Maybe histamines, the stuff of allergic responses. Red wine, white wine, sake, and beer all contain histamine, though studies go back and forth on which drink has the most. Some say it's red wine, but others don't find much difference among the various drinks.

Another possible culprit is tyramine, which seems to come not from yeast fermentation but a second step that uses the bacteria that make lactic acid — pickling microbes, in other words. Some winemakers use them to eat the sharp malic acid that can sometimes result from a fermentation, thereby softening the flavor of some red wines. Usually enzymes break tyramine down before it can do any damage, but some people don't make enough of those, and others are taking high blood pressure medication that inhibits them. For those folks, tyramine can actually cause panic attacks. In the body, it gets converted to another molecule that actually squeezes into the cellular packages that store the excitatory neurotransmitter norepinephrine, pushing the norepinephrine out into the body where it accelerates heart rate and makes people hyper. Even if you make enough enzyme, though, tyramine can cause the same symptoms as acetaldehyde, including flushing and nausea. Some studies — but again, not all — show that neither white nor red wine contain much tyramine; others say red wine does. The scientific jury is out.

In my experience, lots of people who believe that red wine specifically gives them a headache blame sulfites, and it's possible . . . but not likely. In small subsets of the population, sulfites can induce asthmatic responses and even headache, and as a possible mechanism, sulfites can induce the release of histamine in the body. But red wine actually contains *less* sulfite than white wine.

If you really want to blame a chemical for wine-induced headaches, I might point you to another choice — 5-hydroxytryptamine, more generally known as serotonin. It's a neurotransmitter used widely throughout the brain, involved in things like mood regulation. Psychiatric drugs like Prozac are selective serotonin reuptake inhibitors; they enhance the effects of serotonin in the brain. Red wine induces the release of serotonin more effectively than white wine. Like Prozac, red wine also inhibits its reuptake at synapses in the brain. It blocks serotonin from locking into receptors, specifically a subtype called 5-HT_1. And maybe it's no coincidence that the most widely prescribed and successful class of migraine drugs, the so-called triptans — sold under brand names like Imitrex — attach to that very receptor. And about a third of migraineurs do indeed say that red wine can induce their headaches.

In short: different drinks affect different people differently at different times — even holding ethanol constant. The problem isn't that this fact is a surprise. The problem is that these observations don't quite square with a molecular or neurocircuitry approach. Other drugs of abuse have fairly clear modes of action and predictable behavioral outcomes. But not booze. It doesn't have either.

• • •

In 1969, a couple of anthropologists named Craig MacAndrew and Robert Edgerton put that problem more plainly. Their book, the wonderfully titled *Drunken Comportment: A Social Explanation,* is essentially an ethnological study, rounding up observations on alcohol use patterns from cultures all over the world. MacAndrew and Edgerton were trying to take on the idea that alcohol use universally results in a loss of self-control, that it allows people to break societal norms. In fact, they argued, not only do people hold even more tightly to their cultural rules for alcohol consumption, but those rules (again!) differ from culture to culture.

So, for example, we might ask: does ethanol make people violent? Well, the Papago people of what is today the Arizona-Mexico border region used to get quite drunk drinking a beverage made of fermented saguaro cactus fruit, which was only possible to make once a year, at the end of the dry season. The Papago would have a festival; the men would make speeches, have a party, and pass out. No violence. But when Europeans showed up with whisky, which the Papago drank at times other than the harvest, they started going on "drunken rampages," even against their own families.

Or take the native Tahitians, who refused alcohol when Europeans first brought it as part of their ships' stores, but on subsequent contact displayed all sorts of violently drunken tendencies. But by the time MacAndrew and Edgerton looked back in on the Tahitians, they had widespread alcohol dependence problems, but largely without violence.

MacAndrew and Edgerton describe alcohol's effects on culture after culture, in many cases showing exactly opposite outcomes. Here's one of my favorite anecdotes in the book: The Maori of New Zealand had two kinds of drinking occasions, "sessions" and "parties." Sessions usually took place on a weekend afternoon, lasting several hours, and entailed a bunch of guys lying around, drinking, dozing, and listening to the radio. Parties involved both men and women, took place at night—all night—and usually ended with fights or sex. So: Men-only equals quiet passing out. Men-plus-women equals a rough Saturday night downtown.

The villagers in a place called Taira, in Okinawa, *also* had two different kinds of drinking occasions. When men got together and drank sake after work, their typical friendliness generally gave way to arguments and fights. But when the village threw parties to which both men and women were invited, everyone remained relatively polite. As they drank more, the Taira villagers got happier and more ribald — but never violent or sick.

MacAndrew and Edgerton hypothesized that alcohol's effects existed only, as they put it, within limits drawn by cultural norms. It was only in "confused cultures" like the United States, where alcohol was sometimes prohibited and sometimes lionized, that ethanol-induced behavior became dangerous. MacAndrew and Edgerton walk right up to the line — while never crossing it — of suggesting that ethanol doesn't have any intrinsic effects at all. "Even if, say, the brain physiologists were fully to accomplish their task of explicating the effects of alcohol on the human brain in the most minute and final detail," they wrote, "we would remain no better informed concerning the relationship *between* this now fully explicated state of affairs, and the purportedly resultant changes in man's comportment than we are today."

Drunken Comportment was a landmark work, but it faces all the same flavors of criticism as any work of interdisciplinary social science from the late 1960s. It depends on subjective anthropological reports from observers throughout the late nineteenth and early twentieth centuries, when European scientists didn't always have the best eye for what was really going on in the cultures they were studying. A lot of the researchers were men, spending time with men, which means they sometimes missed how women felt about alcohol. Maybe most importantly, says Robin Room, a sociologist who specializes in alcohol and its effects (and speaks highly of MacAndrew and Edgerton's work), the evidence in the book that in a handful of cultures ethanol has no effect — that there's no difference between drunkenness and sobriety — is really shaky. Subsequent population-scale research has shown that one thing ethanol does, as distinct from any other drug of abuse, is increase levels of violence. It's true that the *markets* for illicit drugs can be violent, but except for some stimulants (like methamphetamine), alcohol is the only drug that has the intrinsic ability to make people violent — across cultures and genders. Nevertheless, MacAndrew and Edgerton's work still defines the gap between the way behaviorists and biologists understand ethanol.

Since Alan Marlatt built his BAR Lab in Seattle, the idea has spread. James MacKillop, a psychologist, has a small one at the University of Georgia that he uses to study addiction through the lens of behavioral economics. The two-way mirrors there are decorated with UGA logos, and subjects use a computer to register their desire for another drink as the virtual price at the bar goes up. "For a while my project coordinator was pregnant, and when she had to go out and buy a lot of alcohol she would get terrible looks," MacKillop says. Will Corbin has a bar lab at Arizona State, aimed more toward helping college students avert drinking problems. (It can't be an accident that universities with reputations for heavy party cultures, sited in towns with intricate, built-out bar neighborhoods, also have bar labs.)

No matter where they're located, though, the subtleties of behavior that bar labs can study are difficult to reconcile with decades of research on cell cultures and laboratory animals. "There's a gap the size of the Grand Canyon between the two," says Fromme, the former Marlatt student. In keeping with the trend toward cute acronyms, Fromme calls her lab Studies on Alcohol, Health, and Risky Activities, or SAHARA, and it's an elaborate simulation. The décor is convincing, with an L-shaped bar, cool lighting, and other sensory cues that signal "drinking establishment." Fromme even has her own delicious placebo cocktail. "You get the olfactory cues. You watch me mix the drink. Voilà, that's an effective placebo. Any feelings they report, any behaviors they engage in, are a function not of alcohol's pharmacological effects but of expectancy."

Besides her lab, UT Austin also has the Waggoner Center for Alcohol and Addiction Research. "I go to their journal groups and seminars sometimes, and it's extremely molecular, biological. We don't talk the same language, we don't read the same journals, we don't use the same methodology," Fromme says. "I believe that the effects of alcohol are never going to be explained in a cell or in an animal, because there are inherently other factors involved."

"That must make for some awkward cocktail parties," I say.

"We just don't talk about work," Fromme answers.

Decades after Marlatt started looking at expectancies, researchers like Fromme have started to understand them a little better. People with positive expectancies for how their drinking will turn out — they'll be more outgoing, have better sex, whatever — tend to get those better outcomes. The converse is true, too. If you expect to become more aggressive or do things you'll regret, you're more likely to get that less fun evening. Which set of expectancies you subscribe to seems to depend on early-life modeling, what you've seen in the media and what kinds of behaviors you recognize and remember in your own parents, if they drank.

Fromme's next set of experiments attempts to bridge the research canyon. For six years, she has been surveying UT students on their drinking behavior with a reliable, self-reported questionnaire. Fromme now has data on behavior from over 2,000 kids starting before they leave home and ending after graduation. Now she'll collect saliva from them and sequence a particular set of genes that seem to affect alcohol intake. When we speak, the first 500 samples have already been sent off for genotyping. "What we are predicting is that a serotonin transporter polymorphism will be related to sedative effects in the lab, and two other genes, GABRA2 and OPRM1, will be related to stimulant effects," Fromme says. Different phenotypes will have different responses to alcohol, and those responses will predict — hypothetically — people's long-term outcomes in terms of risky behavior and addiction.

Eventually, Fromme says, her bar lab work and survey work will overlap with genetics, and with experiments by other researchers using imaging (like Mitchell) or higher blood alcohol levels than Fromme can get in the lab. "I really think the answer is going to come from using all these multiple methodologies, which hopefully will converge," she says. "But I've been at this for thirty years, and it's not like I have it all figured out."

What happens to people when they drink, then — even at moderate, "social" levels — is highly individualized, multifactorial, and dependent on cultural rules and references as well as contextual influences like setting and timing. All of which are the ways scientists say, essentially, "Dunno."

In 1983, the reporter Leonard Gross set out to try to parse a very specific aspect of these questions in yet another delightfully titled book, *How Much Is Too Much? The Effects of Social Drinking.* He didn't exactly get to the bottom of it, but he does recount a lovely riff from, he says, a Mississippi state senator, asked in 1958 how he felt about whisky:

> If, when you say whiskey, you mean the devil's brew, the poison scourge, the bloody monster that defiles innocence, yea, literally takes the bread from the mouths of little children; if you mean the evil drink that topples the Christian man and woman from the pinnacles of righteous, gracious living into the bottomless pit of degredation and despair, shame and helplessness and hopelessness, then certainly I am against it with all my power.
>
> But if, when you say whiskey, you mean the oil of conversation, the philosophic wine, the stuff that is consumed when good fellows get together, that puts a song in their hearts and laughter on their lips and the warm glow of contentment in their eyes; if you mean Christmas cheer; if you mean the stimulating drink that puts the spring in the old gentleman's step on a frosty morning; if you mean the drink that enables a man to magnify his joy, and his happiness and to forget, if only for a little while, life's great tragedies and heartbreaks and sorrows, if you mean that drink, the sale of which pours into our treasuries untold millions of dollars, which are used to provide tender care for our little children, our blind, our deaf, our dumb, our pitiful aged and infirm, to build highways, hospitals, and schools, then certainly I am in favor of it.
>
> This is my stand. I will not retreat from it; I will not compromise.

Eight

HANGOVER

GOOD MORNING, SUNSHINE! You are so screwed.

The light coming in through the window is so . . . *there.* You'd kill for a glass of water and die if it came with food. Your guts are in full rebellion; whatever happens next is going to happen in the bathroom. And for some reason you can't remember how to read the clock next to your bed, even though you used to be able to do it relatively easily, you're sure.

You have at least a couple of the following symptoms: headache, malaise, diarrhea, loss of appetite, the shakes, fatigue, and nausea. You might also be dehydrated, and feel generally slow — a little stupider, a little less coordinated. You, my friend, have a hangover.

Scientists have a more inscrutable name for it: veisalgia, from the Greek word for "pain," *algia*, and *kveis*, a Norwegian word meaning "uneasiness following debauchery." That sounds about right.

It could be worse — a hangover so severe that it causes psychiatric dissociation is called Elpenor syndrome, named for a sailor in the *Odyssey.* He gets drunk the night before Odysseus's crew is scheduled to leave Circe's island and falls asleep on the roof of her castle. Elpenor wakes up amid preparations for departure the next day and, hung-over, falls off the roof and dies. The crew doesn't notice Elpenor is missing and sets sail; Odysseus later runs into him in the underworld, where Elpenor begs him to go back and bury his body in an anonymous grave —because he's embarrassed at the manner of his death. Which might be another familiar feeling.

How familiar? Governments often try to calculate how much money the economy loses due to alcohol use by summing up productivity lost to hangovers, to people who can't make it into work the next day. In the United States, that's $160 billion a year.

Even if we try to confine ourselves to moderate drinkers drinking moderate amounts, sometimes mistakes are made. While 23 percent of people do not get hangovers (the scientific term for them is "jerks"), hangovers affect millions of people, maybe billions, and—here is the amazing part—"what causes hangover? Nobody really knows," says epidemiologist Jonathan Howland. "And what can you do about it? Nobody knows." It wasn't even until the last decade that researchers agreed on a basic definition of a hangover, much less started thinking carefully about how to treat it.

For all that ignorance, though, a few compounds may actually help with the symptoms. And walking those backward, a small group of hangover-focused researchers has even started coming up with some explanations for a mechanism.

Howland, a professor of emergency medicine at the Boston University School of Public Health, mainly studies falls among the elderly. But in the mid-2000s, he and Damaris Rohsenow, an alcohol and drug abuse researcher at Brown University—and collaborator with Alan Marlatt, who put together the first bar lab I talked about in the last chapter—started looking at the consequences of heavy drinking. They were more interested in how hangovers relate to the ability to perform at a job. "We were interested not so much in hangover as a cluster of symptoms, but in impairment the day after heavy drinking," Howland says. "As a function of that, we got interested in hangover, initially as a possible explanation for the impairment."

They found that, aside from a flurry of work from researchers in Scandinavia in the middle of the twentieth century, hangover has been largely ignored by science. Nobody even had a survey instrument that would work in controlled studies, and that could assess the severity of a hangover while it was happening — necessary prerequisites for doing useful research.

Anyone who has ever ordered one final, stupid drink before stumbling into a cab home most likely has a great deal of interest in the underpinnings and treatment of hangover. Yet while entire institutes at the National Institutes of Health dedicate themselves to alcohol and drug abuse, almost none of that work is hangover related. In 2010, one study did the math. As of publication, PubMed, the international database of biomedical journal citations, showed 658,610 cites over the last five decades on the subject of alcohol — presumably studying modes of action, addiction, related diseases, and so on. Hangover got 406 studies. That's it.

Still, a few researchers were nosing around on the same path as Howland and Rohsenow. So in 2009, a researcher in the Netherlands named Joris Verster got everyone together for an informal meeting. They dubbed themselves the Alcohol Hangover Research Group, and they even have a logo. It's a kind of shield-shaped thing with the letters AHRG at the top, poorly kerned, and an image of a tipped-over wineglass with a few drops spilled out. Behind it is a pint of beer in a glass decorated with — prepare to have your mind blown — the AHRG logo in miniature. It's exactly the kind of infinite recursion that would make someone with a hangover vomit.

Over the past couple of years, the AHRG has been able to pin some basics. Depending on height and sex, getting your blood alcohol above 0.10 pretty much guarantees a hangover the next day, with symptoms peaking at about twelve to fourteen hours later. In fact, hangovers are at their worst when BAC is at or near 0. And while some researchers have argued that hangovers constitute a sort of mini-withdrawal syndrome, like what happens when an addict goes cold turkey, that seems to be wrong. Some of the symptoms overlap, but withdrawal comes with high blood pressure and a faster EEG, for example, and hangovers are usually the opposite. Anyway, people undergo withdrawal after they've been drinking for days and then stop; hangovers occur after just one session, and they don't last as long. And you don't see snakes.

Rohsenow and Howland's research turned up the 23 percent number of people who are hangover-resistant. They thought those people might point them toward a genetic underpinning for hangover sensitivity. Hypothesizing that being hangover-proof was related to polymorphisms in the genes that code for alcohol dehydrogenase, Rohsenow and Howland set up the usual methodology: a group of lucky (or unlucky, depending on how you look at it) participants get taken up to 0.12 BAC. They sleep it off in the lab, under the supervision of an EMT. The next morning, the participants fill out the Acute Hangover Scale questionnaire Howland and Rohsenow helped develop.

They were looking at variations in the ADH genes clustered on chromosome 4, looking for single nucleotide polymorphisms — changes in just one of the bases that encode the gene. And they got a bit of a hit. Specific variations in a gene called ADH1C seemed to correlate to a lack of perceived hangover symptoms. The downside: those same variations also correlate to a risk for alcoholism. That fits with the idea that people who are less sensitive to the overall effects of alcohol are also the ones most likely to become dependent on it. But these results are preliminary at best. "It was a very underfunded study, so we could just look at four genes," says Howland. "And we were only able to do those four genes on about a hundred people." Howland's team presented the research at a conference, but never published it for peer review.

The same goes for an attempt to stuff hung-over people into an imaging magnet. It was a small pilot study, unpublished — the researchers found eight people already in the lab for a different hangover study and put them into a functional magnetic resonance imager, scanning which parts of their brains were activated during a standard test of cognition and concentration. The areas of the brain that lit up were so diffuse as to be uninteresting. But while the hung-over group didn't perform any worse than the control group, they used more brain to do it — a larger region of the cortex lit up, in other words. Rohsenow thinks this might be an example of a phenomenon called "compensatory recruitment," in which the brain works harder to produce the same outcomes. People with hangovers might not be impaired, but their brains have to pedal much faster to stay on the same pace.

The AHRG members may not have come up with much positive data in terms of causes or treatments, but they did some good work reviewing the ideas that are already out there. They came to one overwhelming conclusion: pretty much everything anyone has ever told you about the causes of hangover is wrong. Or as Howland would have it, probably more accurately: unproven.

Dehydration? It makes sense, sure. Alcohol suppresses the antidiuretic hormone vasopressin, which ordinarily keeps you from peeing too much. Plus, if you're drinking booze, you're probably not drinking water. But in terms of hangovers, levels of electrolytes don't differ too much from baseline controls — and when they do, they don't correlate with hangover severity. So yes, drinking booze dehydrates you. But that doesn't cause the hangover. Plus, drink a glass of water. Now you're hydrated. Did your hangover go away?

How about acetaldehyde, that toxic byproduct of ethanol breakdown in the body? It's another good guess; a lot of the symptoms of acetaldehyde toxicity overlap with hangover. Unfortunately hangover symptoms are at their worst when acetaldehyde levels are low — and again, those levels don't correlate with severity. You can probably cross it off the list, though to be fair acetaldehyde is tough to study because it tends to evaporate before you can get a good reading.

Blood sugar has some intuitive power behind it. Dehydration causes glucose levels in the blood to drop, and the body compensates by creating other sources of energy. Free fatty acids, ketones, and lactic acid all build up, making the blood more acidic in general. That's called metabolic acidosis, and it, too, has some symptomatic overlap with hangover. Hangovers definitely correlate with low blood sugar, but nobody has been able to show reproducibly that raising blood sugar alleviates them. One interesting study did show that the presence of lactate made hangovers worse . . . and that the only way people reliably had high lactate levels was if they consumed ethanol along with glucose. Time to tell the bartender to hold off on the sugar-rimmed cocktail; maybe those warnings about sugary drinks have something to them, though it's too early to tell for sure. But just as with dehydration, if low blood sugar were the problem, administering glucose and fructose ought to be the solution. And it's not — sugar doesn't help the morning after.

The myth about sugary drinks, though, does get us to the admonitions against drinks with high levels of congeners. Vodka, you might have heard, is supposed to give you less of a hangover than red wine or whisky. There might be some truth to this one. Few researchers have studied the relative toxicities and effects of things like acetone, tannins, or furfural, the congeners that make brown liquor taste like brown liquor. Indeed, one study — dicey, I should say, since it was only presented at a conference and not published — ranked different types of booze in order of hangover severity: brandy, red wine, rum, whisky, white wine, gin, with vodka coming in last.

That's not to say that vodka doesn't cause hangovers, though. A comparison of people who drank enough bourbon or vodka to get to between 0.1 and 0.15 BAC — which is super-drunk, by the way — showed that all of them got hangovers. But the bourbon drinkers reported *worse* hangovers.

If we're looking for congeners to blame, we might set our sights on methanol. Levels aren't high in anything you buy in the store, because the stuff can kill you, but it's present at nontoxic levels in almost every alcoholic beverage. The enzyme alcohol dehydrogenase breaks it down rapidly in the body, but where ADH turns ethanol into acetaldehyde, it turns methanol into formaldehyde. These molecules are toxic and very unpleasant. The science here isn't clear, because some studies dismiss the effects of methanol and its metabolites, but one piece of evidence is suggestive: the relative efficacy of the "hair of the dog" — drinking more booze. Ethanol might help with a hangover because it stops the body from breaking down methanol.

On first sip, methanol gets you just as drunk as ethanol does. Both are, in medical parlance, depressants of the central nervous system.

After consuming a lot of methanol, you might feel fine for a few hours, up to a full day. And then you get sick: vomiting, dizziness, and a variety of flu-like symptoms. That's ADH making formaldehyde, which is bad, but doesn't last long. The problem is, it turns into formic acid — ant venom.

Formic acid or formate, a product of the acid, inhibits the action of an enzyme called cytochrome oxidase, which is vital to a cell's ability to use oxygen. Under normal conditions, the eyes, specifically the optic nerve, use a huge amount of oxygen — that's why a couple of the first signs of suffocation are tunnel vision and the loss of the ability to see color. So with a big enough dose of methanol, the eyes go first. And in fact people killed by methanol show a characteristic pattern of lesions on the optic nerve and in the brain.

Eventually the reduction in cytochrome oxidase activity leads to general neurotoxicity. If you live, you end up with Parkinson's-like tremors, slurred speech, difficulty walking, and trouble thinking.

The thing is, alcohol dehydrogenase would much rather stick to ethanol than methanol — one way doctors treat methanol toxicity is to administer a lot of booze. The enzyme goes to work on the ethanol, so the methanol doesn't turn into formaldehyde, which means no formic acid or formate. The patient just pees the methanol out, or exhales it.

Hair of the dog used to be a venerable remedy. In heavier-drinking eras, like before (and during) Prohibition, morning cocktails had a whole chapter of the bar manual. They were called pick-me-ups, and they include almost any of the egg-containing drinks you'll find on classic menus, like the Ramos Gin Fizz. My favorite is the Corpse Reviver #2, which I talked about last chapter in the section about absinthism. The "corpse" in the name is the poor sucker who overdid it the night before. If someone had been standing next to Elpenor with a Corpse Reviver #2, he might still be alive today. Well, not *today,* but you get what I mean.

Nowadays, not many cocktails are socially acceptable at breakfast. The Mimosa and the Greyhound, champagne with orange or grapefruit juice, work, as do the members of the Bloody Mary family, spicy tomato juice with a base spirit. (Try it with tequila — it's called a Bloody Maria, and it actually tastes good, as opposed to the Bloody Mary, which is just a ruined glass of tomato juice.) Unfortunately, drinking more merely postpones a hangover, and as a behavior it correlates with problem drinking in later life. Which, when you think about it, isn't exactly counterintuitive, even though more than one in ten social drinkers admit to having tried the remedy, such as it is.

The best theory going today about what really causes hangovers is that they are an inflammatory response, like what happens when we get an infection. Hangovers are accompanied by elevated levels of molecules called cytokines, molecules used as communications signals by the immune system. One research team in Korea found elevated levels of interleukin-10, interleukin-12, and interferon gamma in their hungover subjects. If you inject those into a healthy subject, that person will start to have all kinds of familiar-sounding symptoms, including nausea, gastrointestinal distress, headache, chills, and fatigue. Potentially even more interesting, higher-than-normal cytokine levels also mess up memory formation, which might account for ethanol-related lapses in recall as well.

As unpleasant as that sounds, it's actually good news. Because a mechanism for hangovers means researchers have a target for treatment.

Few three-word phrases inspire less confidence than "according to Yelp," but according to Yelp, the long, airy shop I am walking into with my two sons is home to the best Chinese herbalist in the East Bay. It's on the fringe of Oakland's Chinatown, amid the light industry, parking garages, and sketchy cell phone dealers that form an inter-neighborhood adhesive matrix with nearby Jack London Square.

I don't speak Mandarin, so I type the words "oriental raisin tree" into the Web browser on my phone and get the Latin name of the plant, *Hovenia,* and Chinese characters. I show the friendly man behind the counter my screen. "Do you have this?" I ask.

"Oh," he says, smiling. "For alcohol detoxification. How much you want?"

Beats me. What do I know from Chinese herbs? "Enough for, like, four servings?"

The man nods, spreads a square of wax paper onto the glass countertop, and turns to the wall of drawers behind him, each the size of the ones in which libraries once kept card catalogs. He opens one, reaches in, and pulls out a handful of what look like sticks. He dumps them out on the paper and indicates that I should . . . what? Approve?

"Taste," he says. "It's good."

I do. It's fine. Tastes like cinnamon. "What do I do with it?" I ask. "Make tea?"

"Take about a quarter, make tea," the man says. He wraps the oriental raisin tree branches into a square wax-paper bundle, ties it together with string, and hands it over. I owe him less than $5.

On the way back to the car, my seven-year-old says, "Papa, what's that for?"

"It's supposed to make you feel better if you drink too much booze," I say. Saying it out loud makes me feel a little like a fraud. I am an active-ingredient kind of guy, not a holistic-approach-plus-chi kind of guy. And unfortunately for my boy, his question sends me into a long discourse about different cultural approaches to medicine and the relative advantages of the Western style of scientific interrogation. He'll probably end up becoming a psychic energy healer out of spite.

But this is different, I reason, because Hovenia may actually have a verifiable active ingredient: ampelopsin, also known as dihydromyricetin, found in Oriental raisin and part of the traditional Chinese pharmacopoeia. It keeps you from getting drunk, and it cures hangovers. Maybe.

The researcher who clued me into it, Richard Olsen, is a neuroscientist at UCLA. He studies alcohol, specifically within the relevant range of blood alcohol concentrations between 0 and a couple of drinks. In that range, he says, the neural mechanisms that respond to alcohol are very specific, and present very interesting targets for treatment.

Not everybody agrees with Olsen, but he thinks that at lower concentrations the neurotransmitter GABA is the most important one

—specifically, he says, one particular receptor that responds to it. Usually, receptors cluster beneath the business end of another neuron, ready to receive the neurotransmitter when it gets released. But more receptors spread out along the neuron instead of staying at the synapse exactly. "Even though they're less dense, there's a huge number," Olsen says.

Their job is to mop up overage—to respond to really, really big jolts of neurotransmitter, huge signals that overwhelm the postsynaptic receptors. These "extrasynaptic receptors" are also, it seems, exquisitely sensitive to anesthetics and ethanol. "They contain a unique subunit called delta, and delta-GABA-R is our baby," Olsen says. It is, as he puts it, "a unique ethanol receptor that responds to low concentrations of ethanol, as produced by one glass of wine, in the brain."

If Olsen's right, this could be the ethanol target everyone is hunting for. On his side is the fact that a drug that binds to the receptor, a benzodiazepine (like Valium) called RO 154513, blocks the effects of ethanol in rats. Of course, like all benzos, it also knocks you on your ass. (To be fair, different subtypes of GABA-A receptors found in different places in the brain, with different subunit compositions, can respond very differently to benzodiazepines.) Another piece of favorable evidence: after repeated exposure to ethanol, the normal plasticity of the brain kicks in. The neurons start to manifest a slightly different kind of receptor that's less sensitive to ethanol but also less sensitive to GABA—which means all those neurons are more difficult to inhibit. Specific areas of the brain end up overexcitable, leading to tremors, almost a pre-seizure condition. The symptoms look a lot like hangover.

Knowing that they were looking for a drug that would bind to the delta subunit of the extrasynaptic GABA receptor—and nothing else—one of Olsen's postdoctoral students, a researcher named Jing Liang, started experimenting with herbs from her native China, beginning with the ones that traditional medicine claimed had an effect on alcohol. Liang has been sitting with us in the conference room, silent. Now she pipes up: "*Hovenia.* It's been used in Asia for 500 years," she says. "I found it in a grocery store."

"A grocery store?" Olsen says, impressed at her ingenuity. "Good for you."

The lab purified the plant until they had an ingredient that acted on the right receptor. It turned out to be a flavonoid, a common molecular family. It already had a name—ampelopsin—but they started talking about it according to the naming conventions of organic chemistry: dihydromyricetin.

"Jing gave a talk at a meeting about our results, and we invited our friends to the bar afterward to try it out if they wanted to," Olsen says. "Now, this is not publishable, and you can't use it for evidence for the FDA, but it's good for us to know what kind of dose we should be using in our clinical trial—and that it doesn't hurt anybody, and does something to us that we want."

"So what happened?" I ask. "Did it work?"

The people who took the pill all reported feeling less intoxicated than they would ordinarily, he says. And they felt less hung-over the next day. Scientific research conferences are famous for active bar scenes, but I assume that the ones at alcohol research meetings rage the hardest and result in the most guilt afterward.

"Have you tried it?"

"Oh, yeah," Liang says.

"She's not able to drink very much at all," Olsen explains. "Because of the Asian metabolism."

Later, when I email Liang to follow up, she tells me that their backer has started selling dihydromyricetin as an over-the-counter supplement. It's called BluCetin. "I take it two times a day," Liang says. She says she's sleeping a lot better now, too.

In 2012, a Duke-trained anesthesiologist named Jason Burke bought a bus, a 1993 Eagle 15 formerly used as a touring vehicle for a Christian gospel family act. He had it driven from Tennessee to Las Vegas, where he remodeled the interior to include a lounge full of bunk beds, and then had the sides painted with the words "Hangover Heaven."

On board, for about $160, the hangover-afflicted can get an intravenous drip of saline solution spiked with vitamins and antioxidants, with an anti-inflammatory and an antinausea drug also on the menu. "I've always been prone to hangovers. Three glasses of wine and I have a pretty rough next day," Burke says. Mostly he relied on Advil and Gatorade, but when he was in his residency, he started hearing about other residents — not to mention paramedics, nurses, and pretty much anyone else with access — using IV saline. Sometimes they'd even travel with bags, heading to Vegas or wherever with a suitcaseful. "One day I was working in the recovery room, with all these people who had postoperative nausea and vomiting, postoperative headaches. And I'd had a pretty significant hangover the previous weekend," Burke says. "I thought, you know, the same things I use here should work on hangovers."

For sure, some people report good results with IV therapy. A clinic in Chicago offers a similar regime. Burke reports that since opening the bus — he'll come to your hotel room, too — his company has treated over 10,000 people. "I was in a fraternity back in college, at UNC. This was in the days before they did irritating things like check IDs. I studied hard, but I partied hard. And now I realize I was a complete amateur," he says. "We saw some epic hangovers this weekend. We went through fifteen barf bags."

Does it really work? Well, a lot of people put their faith in electrolyte drinks like Gatorade or Pedialyte. Whether they help is an untested hypothesis, but the anti-inflammatory and antinausea meds sound pretty good. On his website, Burke offers some less well-vetted advice. "Spend a little bit more money on purer and clearer alcohol," Burke admonishes, before suggesting you might also want to buy the vitamin supplements he sells — which contain not a single ingredient that has ever been shown to prevent or alleviate hangover. "When drinking in moderation, consume plenty of water to counteract dehydration, eat foods rich in vitamins and nutrients to restore those depleted during alcohol consumption, and sweat out toxins through dancing or other forms of safe, physical activity." It's all the usual mythology, in other words. And the remedies that people cling to, like aspirin, or drinking lots of water, or eating a fatty breakfast? Frustratingly, nobody has subjected them to scientific tests.

What compounds have researchers actually tried? For decades, they've looked for overlap between migraine and hangover. Both often involve headache, as well as symptoms like fatigue and sensitivity to light and sound. In 1983, a group of researchers in Finland took that link even further. They knew that if you administer, to healthy people, high doses of immune-system compounds called prostaglandins, you could reproduce what looked like a hangover right away — headache, flush, nausea, diarrhea, restlessness, the works. And elevated prostaglandins are also a hallmark of an inflammatory response.

So the Finns got a prescription anti-inflammatory called tolfenamic acid. It's a prostaglandin inhibitor that isn't sold in the United States, but overseas doctors prescribe it for migraine under the trade name Clotam. The crazy part: it worked. Their two dozen or so subjects got 200 milligrams of tolfenamic acid, fought their way to a heroic blood alcohol concentration of around 0.2, took 200 milligrams more, and then went to bed. By which I mean they passed out.

Twelve hours later, most of the main symptoms of hangover were greatly alleviated, according to self-report, as compared to the unlucky placebo group. Headache, dry mouth, thirst, vomiting and nausea, fatigue . . . all the numbers were better.

That's good news, too, for another compound with proven effect on hangover symptoms, extract of the skin of the prickly pear cactus, *Opuntia ficus indica.* Mexican restaurants sell the paddles of this plant as *nopales,* and they're delicious with eggs. The plant also seems to induce the body to produce heat shock proteins, protective molecules that repair cellular damage. People who make a lot of them naturally tend to be less affected by altitude sickness, for example — a disorder whose familiar symptoms include headache, nausea, and fatigue. *O. ficus indica* extract also inhibits prostaglandin production, and up against hangovers it, too, alleviates symptoms. And even if it isn't as effective as tolfenamic acid, *O. ficus indica* has one big advantage: it's available over the counter, as an herbal supplement.

A couple of other compounds have had encouraging results. A combination of herbs used in traditional Indian Ayurvedic medicine called Liv.52 looks good, but the studies were conducted by the manufacturer so they're not as trustworthy. The makers of Liv.52 say their mix of powders derived from flowers like Himsra and Arjuna, among other ingredients, accelerates the metabolism of ethanol in the liver. Unlike pulling dihydromyricetin from *Hovenia,* nobody has isolated an active ingredient from Liv.52, or connected it to a mechanism.

Burke offers a vitamin cocktail in his IV bags, but the only vitamin that anyone has shown to have an effect on hangover symptoms is an analogue of B_6 with a slightly different structure, called pyritinol —it's two B_6 molecules stuck together. Nobody knows why it might work.

All that is to say that Burke's hangover cocktail has a much better chance of working than, say, two raw eggs in a shot of bourbon. It probably also helps that his company administers it in calming surroundings. "I bought the Eagle specifically because it has a very smooth ride," he says. "Can't be bouncing people around with hangovers." It's a thoughtful touch.

The anti-inflammatory Clotam, the vitamin B_6 analogue pyritinol, Ayurvedic herbal compound Liv.52, and *Opuntia ficus indica* extract are the only four medicines or supplements that actual clinical trials have shown to be at all effective in treating hangovers. Add dihydromyricetin to the list—the thing Olsen isolated—even though it hasn't had rigorous human trials. This is a chunk of information so vital that it demands a stunt.

I invite a couple of hard-drinking friends over with promises of the most exotic, expensive booze I own and cab rides home—if they promise to get drunk and try my hangover remedies. I collect everything but the Clotam, ask my friend Rob to bring his Breathalyzer to make sure we all get above 0.1 BAC, and we get to work.

Rob mostly sticks to tequila on ice with a lime—he's on a high-protein, no-sugar diet. But nothing breaks a dieter's will power like ethanol; by drink number four he's willing to try a Mai Tai—two different kinds of rum, orgeat, and curaçao. Eric, a protein chemist by trade, starts to systematically work his way through my single malt Scotch collection. As for me, I start with whisky, because it looks good when Eric has a glass. I also make a Vesper, the cocktail Ian Fleming invented for James Bond — gin, vodka, lemon, and in Fleming's version, a bitter quinquina called Kina Lillet. You can't get Kina Lillet anymore; I use Cocchi Americano.

The remedies I've been able to get ahold of are stacked on my coffee table. As an icebreaker, I walk Rob and Eric through everything I've already said in this chapter. The two drinks we've all had so far either make them extremely interested in what I'm talking about or make me think I am being extremely interesting.

I'll admit that the experiment was not scientifically rigorous. No control group, for one thing. I asked Rob and Eric to merely report whether, upon trying the supplements, their hangovers felt not as severe as they might have otherwise.

We run into trouble almost immediately. After some large number of drinks, Rob takes out his Breathalyzer, one of several he acquired while reporting a story. It's battery powered, and we put fresh batteries in, but it won't calibrate. Rob, following directions written on the side, blows into it until it clicks, but the readout says something like 0.4, which is an emergency-room blood alcohol content. Now we have no way of knowing if we've reached the magic 0.1 BAC level.

For some reason, that makes it seem like we have to be *absolutely sure* that we are drunk enough to be hung-over the next day. I go make us all another round. The German digestif Underberg, which comes in tiny, paper-wrapped bottles, becomes involved at some point. My notes say I make a cocktail called a Widow's Kiss — calvados, Chartreuse, and Bénédictine. Ordinarily it's delicious, though I do not remember making it or drinking it. At this point my notes become illegible. The outcome is assured.

I give Rob the pyritinol — without telling him that I thought it actually helped me after three drinks the previous night. Because Eric is of Asian descent and has suggested he might have the classic sensitivity to acetaldehyde, I give him the Liv.52. (I don't tell him that the human studies on it are the weakest in the literature, because that would bias the study.) I tell them both to take one dose right then, and then another when they wake up.

I take out my phone, which at this point seems like utterly unfamiliar technology, like something aliens have left in my pocket, but I manage to summon a car for Rob via Uber. It shows up, but the driver stops around the corner, so I walk Rob to the car in my slippers. Eric's fiancée, who has only had two drinks, scoops Eric out of his chair, kneads him into a ball, and squishes him into their car — at least, I assume that's how they got home. They weren't in my living room the next day.

I vaguely realize that I didn't give anyone the cactus extract, but can't figure out how to solve the problem. After a dihydromyricetin tab, I flee to bed.

The next morning is very, very terrible. Typically my worst hangovers sit heavily in my guts, with horrible nausea the main symptom (plus others I won't trouble you with). I get foggy, too — like, can't remember how to type. Today I add a suite of symptoms so migraine-like I suspect the booze might have induced one. Even the attenuated sunlight of an overcast day is painful, and my forehead feels like it has a railroad spike embedded in it. Desperate, I parcel out one capsule each of Liv.52, pyritinol, and *O. ficus indica* extract, force them down with a half swallow of water — all I can manage — and then quietly try to die. I'm not steady enough to get out of bed until 3 P.M.

That's when I finally manage to check my email. Eric, it turns out, threw up at 2:30 A.M. "Yeah, I didn't enjoy the aftermath very much," he says. "Woke up around 7 A.M. feeling generally OK, with a fading, very mild headache. I would say this pattern is typical when I drink past a certain threshold, which I've only done a few times over the past five years or so. Took another of your pills on waking but can't say it seemed to make any difference." He said he seemed to sleep better than he might have otherwise.

Rob reported slightly better results. "I took the pills, which I will say seemed to do something. If anything I would describe it as having a bit more energy than I would normally have in that state. And this extra jolt seemed to give me the ability to stay in a decent mood despite feeling like utter crap," he says. Symptoms were gastrointestinal distress, vertigo, and fatigue/confusion. "I felt typically shitty, but somehow more awake. So I would say it helped slightly. If it was just a slight hangover, it might have made a real difference. But this was a full nuke-the-site-from-orbit-style head thump, so it really didn't stand a chance."

My guess here is that the study design was even worse than I first thought. We consumed too much alcohol for anything we tried to make much of a dent in. This wasn't just heavy drinking—it was a binge, the stuff of research on college drinking behavior. It was the exact thing I've been saying I didn't want to get into in this book, in fact. Not healthy. The ethanol might not have even metabolized out of our systems by the time we woke up.

On the other hand, I rationalized, my N=3, control-group-free study was almost as good as most of what already exists in the literature. Even the critical papers, like the ones on *O. ficus indica* and Clotam, have small group sizes. Pharmaceutical companies haven't wanted to touch therapeutics for some reason—even though it seems like a cash cow. Burke figures that the Hangover Heaven bus is a boon to the Vegas economy, because it keeps people out in the casinos and restaurants instead of in their hotel rooms curled into a fetal position. He says his team administered treatments to a group staying in a high-roller suite at the Wynn who were so sick they planned to leave, but after their IV fluids they called the pilot of their private jet and told him to get a hotel room to stay another night. This is the kind of story people tell about Las Vegas. The morning after my ill-advised experiment, I have to admit I see the attraction of Burke's method. When you're grasping at straws, by definition your attitude is, "Oh, thank God, there are some straws here! Gimme!"

So why isn't the science better for something with a huge potential market? Shouldn't a smash-hit lifestyle drug be buried somewhere in here? "I think that it's failed to get a lot of attention because I'm not sure what the public health implications are," Howland says. Maybe the problem is that pharma companies and the government worry that a hangover drug would suborn drinking to excess. "I hear that a lot. It's part of the whole American attitude to alcohol, which has this moralistic edge to it," Howland says. "Like getting rid of the punishment would somehow open the door to more mischief." Then again, even Howland hasn't done research on cures. A few years ago he tried to get a project together to look at a commercial hangover cure, but couldn't get the protocol past his Institutional Review Board because the product didn't have FDA approval and the maker wasn't interested in getting it. Beyond that, he says, "I haven't known where to start."

Burke says he's going to use the profits from his Hangover Heaven business to start an institute — he has a friend who's a biochemist, he says — and try to figure out if rehydration really does jump-start metabolism, and whether B vitamins do help with the manufacture of antioxidants in the bloodstream. Meanwhile a mobile hangover spa seems like it has obvious franchise opportunities, right? Mardi Gras? The Austin-based music and media conference South by Southwest? San Francisco's Mission district on a Sunday morning? "We're talking to the transportation authorities and medical boards in various states," Burke says. "The hardest part of the business is licensing. And logistics, because it never fails that everyone wants to get treated at the same time. Ten o'clock on a Saturday morning, and the phones start to smoke." It's the future of drinking in a nutshell: making an appointment with your hangover.

CONCLUSION

The creation myth of Starbucks, the ubiquitous urban coffeehouse, says that CEO Howard Schultz returned to the United States from a trip to Italy — and a lot of visits to little espresso joints — with a vision of a "third place" between work and home, a place comfortable enough for recreation or business that was easy to go to and spend time and money. Obviously it worked. There are a gazillion Starbucks.

So successful was Schultz's idea, in fact, that now we don't notice how many different things go on in any given Starbucks (or Peet's or your neighborhood one-off artisanal coffee place). At any given time of day, people are meeting socially, reading recreationally, or banging on a laptop keyboard, head down over code or a movie script or possibly even a book about the science of booze. The concept of the "third place" shouldn't have taken anyone by surprise, though. Because before Starbucks' ascendancy, the third place was a bar.

The bar, the pub, the drinking house, the gin joint, the tavern — go back as far into history as you want and the place where alcohol is served (and, depending on location and era, where prostitution and gambling were also on offer) has been a place outside the cultural rules governing both private and public behavior. The labels researchers have put on bars — "liminal space," "time-out," "alternative reality," and so on — are the same kinds of terms we use when we talk about alcohol and drinking itself. As a report from the Social Issues Research Centre in England put it, "the drinking-place is the physical manifestation of the cultural meanings and roles of alcohol."

Sociologists who study American drinking habits like to characterize the United States as being more conflicted about alcohol behaviors than more enlightened places like the southern Mediterranean, where, hey, man, a glass of wine is just integrated into everyday life and, like, you know, in France, kids get a glass of wine with breakfast and nobody is uptight about it. At least, that's the fantasy. Taken to its logical extension, alcohol becomes just another recreational drug, one that should be no more or less regulated than any of the others — marijuana, opiates, methamphetamines, hallucinogens — unless it can be shown to cause specific harms. In fact, David Nutt, once one of the British government's top advisors on the harms of drugs, got fired for suggesting that very thing — that if you add everything up, alcohol is actually much more harmful than marijuana. Leave out the economic costs of people too hung-over to work — in 2010 drunk driving caused over 13,000 deaths in the United States and cost over $37 billion. The costs of treating alcohol-related diseases, including addiction, can top out over $20 billion a year. According to one review of emergency room intake records, having just two drinks basically doubles your chances of incurring some kind of injury. So why regulate marijuana so heavily and alcohol barely at all? The laws covering alcohol and recreational drugs in general, Nutt argued, were nonsense. And the British government canned him for it.

American culture is much more likely to wall off transgressions, even mild ones like drinking, in space rather than time. Most of the world has Carnival, where once a year everyone goes nuts on the carnal-sin checklist, and then after a few days things settle back down. Many of the cultures in MacAndrew and Edgerton's *Drunken Comportment* were the same. In the United States, we confine that sort of abandon by location — Las Vegas, New Orleans, the drinking neighborhoods of Austin and Athens, Georgia . . . and, in almost any city, the bar. It is a cross-cultural place-out-of-space-and-time, a space for drinking and, more importantly, punctuation. We drink, often, to note beginnings and endings: the first date, the toast good-bye, the fortifying shot of courage, the after-work bonding experience, the backdrop for romantic exploration — these are the things that bars do. And maybe even more. Based on surveys of bartenders at Veterans of Foreign

Wars meeting halls, a trio of researchers from Ohio State University in 2010 concluded that bartenders were potentially one of the most effective ways to get mental health information and assistance to vets with PTSD. The bartenders, most of whom were long-term employees, described the vets they served as "family."

These cultural frameworks can evolve; increasing legalization of marijuana could lead to acceptable use in public, and that could conceivably change the vibe of a bar — assuming people become willing to share a joint outside in the same way they had to become accustomed to bumming cigarettes off one another when smoking got prohibited indoors.

The processes for how we make and understand drinks with ethanol in them might evolve, as well. The new yeast strains Australian biologist Isak Pretorius was engineering to produce specific effects and flavors stand a chance to usher in a world where all the commodity ingredients of a beer or wine are specially designed. Sylvie Dequin, a yeast researcher in France, is using genetic modification in the laboratory to create strains of yeast that make wine with decent flavors but much lower alcohol levels. She can already reduce the alcohol content by a couple percentage points. Of course, in Europe, the GM strains are nonstarters. Dequin is working on traditionally bred versions that express the same genes.

Sean Myles, the genomicist at Dalhousie University, may never get acceptance for a purpose-built grape strain, but as the climate changes over the next decades, different grapes are going to become a necessity — as are different regions to grow them. Napa, Sonoma, and the Loire may give way to new wine-growing regions. The entire Central Valley of California could be off limits to grapes, as might much of the Sonoma and Napa Valleys and most of Southern Europe. On the other hand, huge chunks of Washington seem likely to open up. The mountains of central China seem like a good possibility, too — if the Chinese can find somewhere else to put those pesky giant pandas.

While the science and technology for fermenting and distilling hasn't really changed in a couple thousand years, it's reasonable to predict the same pace of improvements and tweaks — larger-scale production, maybe, with more consistency of product. If legal trends in craft distilling continue along the same arc as craft brewing did three decades ago, expect more small distillers to start up and be allowed to build their own tasting rooms. They might even move beyond the steampunk technology that powers them today. The English distiller Sacred Spirits is actually one guy, Ian Hart, working from a homemade reduced-pressure still that can separate and recombine flavors from the macerations Hart makes of various botanicals. It's an elaborate version of the rotary evaporation still that Dave Arnold uses at the New York bar Booker and Dax, and it presages a whole new approach for distilleries.

Maybe, even further in the future, something will replace ethanol altogether. David Nutt has been reporting for almost a decade about experiments on a chemical analog to ethanol, an alcohol replacement that would have the same effects—that would act on the same subunits of GABA receptors in the brain, in fact. But it'd be reversible, with an antidote that would instantly sober up a user, or cure a hangover. Nutt says he has five candidate compounds, with their antidotes, ready to go. "After exploring one possible compound I was quite relaxed and sleepily inebriated for an hour or so, then within minutes of taking the antidote I was up giving a lecture with no impairment whatsoever," Nutt wrote in a commentary for the *Guardian* in late 2013. All he needs, he says, is funding to further test and refine the compounds. The alcohol industry might sign up—assuming the negative consequences of drinking on society get serious enough that they have to pony up for an alternative, like e-cigarettes and the tobacco industry. Could he actually have come up with synthohol, straight out of *Star Trek*? Alcohol without consequences? It's a tantalizing promise.

But none of those changes will break the human connection to alcohol across deep time. Our history with the stuff is our history on earth, a history of humans becoming modern, tool-using, technology-making creatures.

In a wide office at St. George Spirits, upstairs from the distilling floor and almost directly above the lab, a group of visitors in leather armchairs admires Lance Winters's books. Next to bottles of Nikka whisky from Japan and antique bottles of Fernet is what looks to be an original edition, from 1871 or thereabouts, of Pierre Duplais's *Treatise on the Manufacture of Alcoholic Liquors,* the ur-source of information about absinthe. And there's *Henley's Twentieth Century Formulas, Recipes, and Processes,* a collection of trade secrets for making just about anything, from gasoline to soda pop. "If you were to wash ashore on an island with that," Winters says, talking past an unlit Partagas, "you could restart civilization."

One guest, Alexander Rose, completely gets that idea. Rose is the executive director of the Long Now Foundation, a San Francisco–based group established to think about humanity on long time scales — millennia, eons, instead of just years. And right now Rose is preparing to turn the bare-bones display space at the center of the Long Now offices into an upscale lounge. Rose is an engineer, a robot builder by trade, and he figures that the best place to talk about the end of the world is a bar.

Rose has come to Winters because Long Now wants to serve something special there. It should appeal to visitors, sure, but really Rose wants to be able to convince potential donors to pay upwards of $5,000 each for their own personal bottle of something, cradled in a purpose-built harness that will drop down from the ceiling to the bar when they show up for a drink. Long Now HQ is in Fort Mason, a collection of shops and theaters on the water near Fisherman's Wharf that is technically a Federal enclave — state and local laws governing the sales of distilled spirits don't apply. Rose can do whatever he wants without a license.

But exactly what Rose wants to put in the bottles is TBD. "It can be as subtle as a label or extreme as a product that people could buy at a high dollar value," he tells Winters. It has to be about "both the very deep past of how alcohol and civilization have become intertwined, as well as the more difficult but open-ended question of, what are the drinks of the next 10,000 years?"

Winters leans in. He gets it. "It's as much a timeline as it is a drink menu," he says. "Any time you can take alcohol out of the traditional framework of commoditization . . . this is one of those one-in-a-million opportunities."

They talk over a few possibilities. A rice wine, to evoke the early Chinese distillers? A whisky? Rose starts talking about his love for the range of St. George gins — he happens to know a lot about juniper. One of Long Now's big projects is a 10,000-year clock, a massive piece of engineering to be built inside a mountain in western Texas. Using nothing for power except gravity, it's supposed to keep time for ten millennia, into the next iteration of humanity, whatever that is. Originally the clock was going to go inside a different mountain, this one in Nevada. Long Now still owns the site, Rose says, and it's covered in juniper — as well as bristlecone pine, a plant that can live as long as 5,000 years.

Winters's eyes widen. Honestly. Just because something is a cliché doesn't mean it never really happens. "If you were able to send some juniper berries, I could run them through and see what they show," he says. "If there were some way to get some needles from a bristlecone, too — that way you could say you had needles from a five-thousand-year-old tree."

Rose smiles. "How much would you need?"

A couple months later, Rose delivers to St. George five pounds of berries and about a pound of needles taken from fallen trees. (Bristlecone is a protected species; the Long Now team isn't allowed to cut anything down.) Winters had hoped for twice as many berries, but the needles go straight into 100-proof spirit to preserve and extract the flavors.

Two months after that, it's gin. At a small reception at the Long Now offices to show off the designs for the remodel, Rose pours the first samples from one of the custom-made bottles, a modified flask with a spherical bottom and cylindrical neck that he sourced from a scientific glassworks in Emeryville. At the makeshift bar Rose has set out a bowl of juniper berries and a bowl of dried orange peel to provide sensory context; I taste the berries — they're sour-sweet and resinous. The gin is more citrusy and oily than juniper-y, to me.

It's just a half shot of booze, but tasted while standing among pieces of a clock designed to tell time for 10,000 years and a table laid with supermarket cheese and salami, the gin feels connected to something bigger — to ancient ingredients, to the Chinese brewers, to Louis Pasteur, to wild grapevines in the Transcaucasian mountains, to an Alexandrian alchemist and Celtic barrel makers and Caribbean bacteriologists and Scottish master distillers.

Human beings had the technology to make gin a millennium before they understood why it worked, why it tasted the way it does and made us feel the way it does. We're getting better at that now. We've learned about the biochemistry of yeast, the mutable chemistry of sugar, how to take the plants we eat and transform them into drinks and alcohol. The biomechanics of fermenting and the physics of the still aren't mysterious, even though brewers, winemakers, distillers, and researchers have plenty more questions about how to optimize the process of making booze.

Even with the gaps in the booze knowledge database, the people who make the stuff are able to do something a little bit magical. It's a creative act — sometimes industrialized and commercialized and sometimes idiosyncratic and artisanal, but in both cases it's human beings coming together to make a thing that didn't exist before they started working on it. More than that, booze straddles the worlds of objective reality and subjective experience. Those producers make a thing that has a demonstrable, quantifiable effect on our physiology. We perceive it with our senses and it changes our bodies.

It also changes our minds. We all taste booze a little differently, and feel its effects in our own ways. How we see alcohol depends on its depictions in the macro-scale culture but also in what we see as children — whether it's something that creates a bonding experience with parents or the source of dangerous abuse. It can be celebratory, dangerous, and every experience in between.

The perfect bar moment, the consumption of a psychoactive chemical in a space preserved for that ritual, arises from intention. We built the booze and the places to drink it. They weren't here before human beings and we didn't stumble on them by accident. We made them that way. We've cracked open the yeast and other microbes responsible and begun to understand the biology that drives the product. We've begun tuning those bugs and the substrates they work on, shifting from the gross, unpredictable business of domestication and husbandry to the cold precision of genetic engineering.

We humans were making booze before we had science, much less a science of booze. Now that we know more, we have more control over the whole process. That makes it a more viable business, but also helps us understand more deeply the pleasure we take in it as consumers. The science of booze, of fermenting and distilling, doesn't detract from the magic behind it. Just the opposite, in fact. To paraphrase the science fiction writer Arthur C. Clarke, magic is really just advanced technology. Science is how we *make* magic.

That Long Now gin doesn't only taste like juniper and orange. It tastes like civilization.

AFTERWORD

The research was not as much fun as you think. At least, not for the reasons you think.

Don't get me wrong—I enjoyed writing *Proof.* Book-writing isn't relaxing, but it's an engaging challenge. The thing is, when the subject is booze, everyone assumes writing also involves a lot of drinking. When I tell people that I have written a book about the science of alcohol, they almost inevitably ask me about "research" in the same tone as you might ask someone about a trip to Las Vegas—as if most of my reporting took place in bars and distilleries (which, OK, a lot of it did, but a lot of it did not) and as if I typed it all with a glass of adult beverage next to the keyboard. I did not—I found no surer way to lose inspiration than to take more than two sips of anything alcoholic. Let me just activate buzzkill mode, push up my metaphorical glasses, and say that even Ernest Hemmingway wrote sober. (He started drinking in the afternoon.) If William Faulkner, Dashiell Hammett, and Raymond Chandler couldn't write while smashed, I certainly can't. And don't.

But I'll admit to a secret desire I did harbor. I always wanted to be able to walk into a bar—more than one, if possible—and know the bartender, be greeted as a regular or at least a trusted repeat customer. And now I have that. *Proof* made me part of a delightful class of people I think of as "professional drinkers"—bartenders, liquor store owners, distillers, brewers, winemakers, and booze writers. Jennifer Colliau, the bartender at San Francisco's Interval and owner of cocktail ingredients company Small Hand Foods, has shared a flight of apple brandies in my dining room. The great booze writer Wayne Curtis and I have comoderated a panel at Tales of the Cocktail, the annual New Orleans gathering that is the San Diego Comic Con of booze (or perhaps SDCC is the Tales of the Cocktail of nerds). On that panel Wayne and I had distillers Lance Winters and Maggie Campbell pour a couple hundred people samples of stuff they made special for the conference — things no one will ever be able to taste anywhere else. I have sampled five-decade-old Highland Park whisky next to Christopher Null of the invaluable Drinkhacker website, and I have tasted strange, experimental beer at Cervecería de MateVeza with William Bostwick, author of the lovely beer history *The Brewer's Tale.* I've talked shop with Jim Meehan, whose *PDT Book* is still my go-to cocktail manual, and oohed and aahed over mad genius Dave Arnold's booze book with Jim Morganthaler, who runs the innovative Portland bar Clyde Common, basically invented the idea of the barrel-aged cocktail, and whose own *Bar Book* has become another standby in my kitchen. Getting to hang with the professional drinkers was like finding my tribe, like a Dickensian orphan learning he has a family after all. I worried that they'd think my reporting and writing were either naïve or prosaic, but they welcomed me as a kindred spirit.

In a sense, then, writing *Proof* has made me a booze geek's booze geek. It's a role I'm happy to play, because right now is an exciting time — even, I'll say, a weird time — for the science that goes into beer, wine, and spirits. This book was in some ways out of date the moment I locked copy, because the progress of that science did not stop. Rudely, it didn't even bother to slow down.

Take, for example, the case of Templeton rye. If you look at a bottle of Templeton — a very tasty American rye whisky — you'd probably think it came from one of the hundreds of small craft distillers that have opened in the United States in the past decade. It's not a crowded field yet, but the number of smaller makers doubles every three years or so, a curve that parallels the growth of small breweries in the 1980s.

Templeton isn't a craft distillery, though. It's what whisky writer Chuck Cowdery calls a "non-distiller producer." He defined that phrase in a May 2013 blog post as a company that buys aged whisky in bulk from a large distillery and slaps a new label on it. As Cowdery puts it, "Virtually all of America's whiskey is made at 13 distilleries owned by eight companies. All of the NDP whiskey is made by those producers too." Templeton is, as some less generous writers have put it, a Potemkin distillery.

In fact, a significant portion of rye available today — Templeton included — comes from an industrial distillery called MGP, in Indiana. While professional drinkers had known about MGP rye for some time, the drinking public was largely unaware the place existed until 2014, when journalist Eric Felten wrote about it in the *Daily Beast.* Today Templeton is the subject of a class-action lawsuit filed for, essentially, false advertising. Buyers, the lawsuit asserts, thought they were getting something artisanal and locavore, when in fact they were drinking something bulk-produced.

Professional drinkers, though, see non-distiller production as something worse than a misleading advertising strategy. To them, it's an existential threat to a nascent industry. Even when non-distiller producers sell perfectly good-tasting juice, goes the logic, they undercut the "real" craft distillers. By producing in bulk something that looks to reasonable eyes like a craft product but costs far less, the NDPs use market forces to guide consumer dollars away from small distilleries at a critical moment in their development, just as they are beginning to flourish. It'd be as if a giant transnational brewer like Molson Coors or Anheuser-Busch InBev made and marketed a craft-like beer without openly acknowledging its provenance (which they did and do, and faced much the same criticisms).

I'd argue that the kerfuffle over Templeton is really about something more important: *authenticity.* That's because Templeton took its deception one step further than the other non-distiller producers. Its owners created a backstory for the brand that at least implicitly advanced a narrative suggesting the rye was made, stills-to-bottles, in Templeton, Iowa, from a family recipe dating back to the days of Prohibition. The label and story behind Templeton gave the rye what the science fiction writer Philip K. Dick called "historicity," or the sensation of being historical, of having historical significance. In fact — as the lawsuit forced cofounder Keith Kerkhoff to admit—Templeton rye was an MGP ready-made recipe to which the company added other whiskies in order to augment the flavor.

In an interview with a big-shot whisky podcaster, and later in a public statement, Templeton's owners explained that they were doing more than just blending, bottling, and labeling. Templeton had also worked with a Louisville company called Clarendon Flavors to develop an additive that gave their booze the flavor they were looking for—something closer to what that old family recipe might have tasted like, though nobody at the company has characterized it any more specifically than that. In other words: Templeton rye is rye with synthetic rye flavor added to make it rye-ier.

This was a blockbuster. Professional drinkers had long passed rumors to each other about flavor and color additives, but this was the first time a company had ever admitted to using them. Templeton's owners, by the way, declined to talk to me on the record, citing the ongoing lawsuit.

Now, if you still want to drink Templeton rye, I figure you're covered by bartender-client confidentiality. It still tastes good, after all. But what's fascinating about the Clarendon additive is that it shows you can chemically synthesize a flavor that's supposed to be unique to distilled rye. I argue in *Proof* that one of the things that made distilling a significant technology is that nobody could simulate the product. If you want something to taste like a distillate, you have to distill it. Yet here we are, finding out that theoretically you could take a neutral spirit or boring white rum, add a flavor, and end up with booze close enough to whisky that it'd be drinkable.

Whether you could then label it as "whisky" without running into trouble with federal regulators and professional drinkers would be another issue. But let's put aside questions of ethics or legality and chew on a more fundamental and interesting problem: Is Templeton, with its synthetic flavor and industrial origins, really *rye?* If it isn't, what should we call it? A "rye-like drink"? And if Templeton *is* rye whisky—by dint of the philosophical qualia of taste, aroma, color, mouthfeel, and so on—then what do we really mean by "rye" at all?

If you're willing to get more theoretical, science undermines the semantics of booze even further. In 2007 Thomas Hofmann, a chemist at the Technical University of Munich, started deconstructing red wine. In a series of papers Hofmann describes using high-performance liquid chromatography–mass spectrometry/mass spectroscopy, ultrafiltration, and other analytical methods to break apart Amarone, Bordeaux, and Dornfelder varietals into their constituent molecules. He identifies the ones that create astringency — the puckering, inside-the-cheek, sucking kind, as well as the bitter type and the velvety mouthfeel type. He also isolates the molecules that create sweetness, and a bunch that contribute to saltiness and sourness. Over the past half-dozen years Hofmann's lab has characterized dozens of compounds in wine according to what they contribute to our drinking experience.

Hofmann's team also tests the molecules they've analyzed by mixing them together and seeing if they produce the correct flavors and textures. He takes the right molecules off the shelf, dissolves them in water and ethanol, gives the solution to a panel of trained tasters, and asks, basically, "Is this like wine?" Using just thirty-seven compounds, Hofmann is able to create what he calls a "taste recombinate" that his panelists say matches attributes of Amarone, a famously complex and deep wine. He then removes the molecules one by one to ensure each one contributes the qualities his hypothesis predicted. His papers suggest that he's especially good at nailing astringency and bitterness.

How scared should the winemakers of Napa and the Loire be of Hofmann's recombinates? Not very, I'd imagine. What works in a red-light-lit tasting room full of panelists who've been training for two years might not fare as well in the wine section at Trader Joe's. Anyway, science already offers winemakers plenty of chemical and synthetic tricks to produce a cheap, appealing, and consistent product. Powdered tannins add astringency. Oak chips or sawdust in tea bags round out the flavor of age. A concentrated grape syrup called Mega Purple deepens a wine's hue. Filtration technologies tune the concentrations of dozens of flavor compounds.

But none of that is building a bottle from scratch, additively. No winemaker is suggesting, at least publicly, shifting from a model of harvesting and juicing grapes for fermentation to combining ethanol, water, color, and a few drops of chemicals and calling it "wine." As far as I know, no wine company has proposed making a batch of flavor syrup and trucking it to local bottling companies. But research is starting to suggest that they *could.* It wouldn't be authentic. It wouldn't have historicity. But it might still be wine — or at least wine-like.

That's a freaky idea. At a beer chemistry conference in Portland, Oregon, I heard a representative from an enzymes company suggest the syrup-plus-local-water franchise model — the approach that makes quality-controlled Coca-Cola available just about everywhere on earth — could work for beer, too. The assembled brewers reacted as if someone had just sent a few volts of current through their seats. It wasn't pretty. It was, however, economically viable — assuming brewers would be willing to figure out how to differentiate a microbrewed beer made locally according to the half-millennium-old German beer-purity laws from something shipped from the Central Office.

I don't think it'll happen. I can imagine transnational drinks companies coming up with ever more innovative beverages containing alcohol — the recent popularity of bourbon with flavors like honey and cinnamon speaks to the skill of large distillers' product-development labs. But the combination of historicity and authenticity with taste and smell is still powerful, because it sums to emotion and memory. If people responded positively to *Proof,* I think it was because the book is fundamentally about how we perceive the world around us — through the mostly familiar experience of wine with dinner, a cocktail at a bar, a beer at a baseball stadium. *Proof* was my attempt to enrich memories with facts, to start with what happens and add the why of the thing. Inventing science-booze wouldn't just disrupt the authenticity side of the equation; it would break the experiential part, too. Booze just wouldn't be as much fun.

Of course, if you're a professional, you don't get to have fun all the time. Philosophizing about the future of the business is all well and good, but the fact is that when I talk to people about the book, after the inevitable impugning of my level of sobriety, most of them have practical, down-to-earth questions. What they wanted to know has taught me, I hope, how to be a *better* professional drinker.

When I was writing I used to come home and ask my wife if what I was working on was at all interesting. I'd blow through the door and, before saying hello or taking off my coat, say things like, "Do people know that whisky is distilled beer and brandy is distilled wine?"

"No," she'd say, usually without looking up. "Nobody knows that but you."

"OK. Good. It's going in the book."

Before I started typing, I promised my editor that I'd do very little history. All state-of-the-art science and the newest research, I swore. But I couldn't stick to it. The history of booze—how people learned about yeast and enzymes, when people first started fermenting, who invented distilling, the early years of studying alcohol's effects on the body—all turned out to be the most effective ways of conveying the science. In chapter after chapter (and in conversation after conversation, lecture after lecture) I have returned to history as an explanatory device, because telling the story of how those people back then—whoever they were and whenever they were—figured out the answers to their questions turns out to be a very good way to answer our questions right now.

And the question everyone most wanted the answer to? The one thing that turned out to be the most resonant, the biggest crowd pleaser? Hangovers. I'll be candid with you: I did that on purpose. The hangover chapter fits the rest of the book only obliquely; every other chapter mostly takes place somewhere between one drink and two, at the level of a slight buzz. And then at the end? Bang. We are blotto together. But I knew people would be desperate for answers. I knew that if a science-of-booze book crossed my desk as an editor, and I hadn't written it, I would look for hangovers in the table of contents.

So it was a real bummer that the science is so scant. My posture in the chapter is one of dismay: Why isn't there more research to write about? Why do I have to talk about folk cures and anecdote?

That made for some awkward silences in interviews about the book,

I'll admit. My worst moment came talking to a reporter for a high-circulation magazine aimed at young women. It took me about twenty minutes before I understood that what she was really looking for was a set of tips for how to go out with friends and drink a lot — like, a whole lot — and not be hung-over the next day. What I was thinking could only sound patronizing and creepy. "Sweetie, you cannot go round-for-round with a Wall Street douchebag who has fifty pounds on you" is what I was thinking. And also: "If that's what your friends are pressuring you to do, then they are not real friends." I didn't say any of that. But I thought it. Hard.

What I suggested to the reporter was that she should adopt what I had by then adopted as a rule-set for drinking professionally as a professional drinker. Limit your rounds. Consider ordering soda water with bitters between the hard stuff. Order good, expensive drinks to enjoy more, and more slowly. Try the weirdest thing behind the bar. Drink whatever it is you have not tasted.

She dismissed it all. That was not the way she wanted to drink, and assuming she stays safe, that has to be OK, too. I may still be the most annoying person in the world to have a drink with, but I'm trying to learn to be a little less judgy. You want to order something sweet and blue, you go ahead. Talking to people about *Proof* taught me a valuable lesson: Sometimes the most important skill a professional drinker can have is knowing when to step aside and let someone else order a round.

ACKNOWLEDGMENTS

The word "actually" is why people don't have drinks with me anymore. For the last three years, every time I was in a bar, or having a glass of wine or a beer with friends, or making small talk at a party, someone would start talking about the drink in his or her hand—about its contents, how it was made, what was in it, its history. And I would start a sentence with "Actually—"

And that was it. My head full of booze data had turned me into the snottiest of snotty bar know-it-alls. So I apologize. The next round is on me.

Even though that antisocial behavior and this book are entirely my own responsibility, I have had invaluable help and input from many people. In fact, a trio of America's top science writers—Carl Zimmer, Bill Wasik, and Thomas Goetz—realized at one critical dinner that there was a book in the science of booze. I'm really glad they pointed it out to me.

Speaking of colleagues, my coworkers at *Wired* have been extraordinarily patient with my attention to this project. The previous editor in chief, Chris Anderson, was gracious with advice on the mechanics of the book business, and both he and current EIC Scott Dadich were forgiving as I did my reporting, sometimes (I must admit) at the expense of work for the magazine. My friends at *Wired* picked up a lot of slack for me, especially Jason Tanz, Robert Capps, Mark Robinson, Caitlin Roper, Peter Rubin, Jon Eilenberg, and Sarah Fallon. I'm grateful, too, for the valuable input of a few readers—Christian Thompson, Mark McClusky, and Daniel McGinn.

Joan Bennett's research on Jokichi Takamine and the history of koji turned out to be a linchpin, as did James Scott's work on *Baudoinia,* the whisky fungus. Both gave more of their time than they had to. James MacKillop even sent pictures and video of his bar lab at the University of Georgia. Above and beyond. Brendan Koerner's understanding of the state of addiction science did a lot to shape my thinking about that field, and Jeff O'Brien and Kate Bottrell offered sage advice on navigating the world of winemakers. Also, a certain friend of a friend loaned me the academic credentials I needed to paw around in a university's online archive of scientific journal articles. How did people do this before computers? Jeez.

To the extent I stayed centered and on point throughout this process, Matt Bai gave me the advice that made that happen. David Dobbs wrote me a memo on finding an agent as brilliantly reported as his magazine work. As proof, it led me to a hell of a representative, Eric Lupfer at WME.

My editor, Courtney Young, managed to not tsk-tsk too loudly at my first draft and guided me calmly and correctly to the right path. Erik Malinowski, a great reporter and writer but also the single best fact checker I have ever worked with, saved me from myself uncountable times.

As much as I might wish I could blame the errors that remain on them, everything wrong is all my fault. I've corrected a few small errors of fact and interpretation in this edition, most of which came to my attention thanks to eagle-eyed readers. So I owe them a thank-you here, too. If you want to see the specifics of what I changed, please take a look at adam-rogers.net/corrections.

Even with the help, kindness, and expertise of all those people, though, this book would not exist without the brain and insight of my wife, Melissa Bottrell. She is my watchword and my beacon. Actually, I love her.

NOTES

INTRODUCTION

page

9 *A single serving of beer:* David R. Antonow and Craig J. McClain, "Nutrition and Alcoholism," in *Alcohol and the Brain: Chronic Effects,* ed. Ralph Tarter and David Van Thiel (New York: Plenum Medical Book Company, 1985), 82.
daily caloric input from ethanol: Benjamin Buemann and Arne Astrup, "How Does the Body Deal with Energy from Alcohol?" *Nutrition* 17 (2001): 638.

11 *Peat, the partially decomposed mix:* "Peat and Its Products," *Illustrated Magazine of Art* 1 (1953): 374.
hydrogen bonds between the two ingredients: Naiping Hu et al., "Structurability: A Collective Measure of the Structural Differences in Vodkas," *Journal of Agricultural and Food Chemistry* 58 (2010): 7394.

1. YEAST

15 *"vanishes in a puff of logic":* Douglas Adams, *The Hitchhiker's Guide to the Galaxy* (New York: Harmony Books, 1979), 54–55.

16 *"and loves to see us happy":* Walter Isaacson, *Benjamin Franklin: An American Life* (New York: Simon & Schuster, 2003), 374.
that provide strength and protection: David J. Adams, "Fungal Cell Wall Chitinases and Glucanases," *Microbiology* 150 (2004): 2029.
That was in 1996: A. Goffeau et al., "Life with 6000 Genes," *Science* 274 (1996): 546.
"but a poor model for other fungi": Jason E. Stajich et al., "The Fungi," *Current Biology* 19 (2009): R843–44.

20 *vinegar was akin to death:* David Dressler and Huntington Porter, *Discovering Enzymes* (New York: Scientific American Library, 1991), 24.
the only ingredients be barley, water, and hops: Ian Hornsey, *A History of Beer and Brewing* (Cambridge: Royal Society of Chemistry, 2004), 321.

help the next fermentation work, too: Dressler and Porter, *Discovering Enzymes,* 23.

the Greek word for boiling: Herman Jan Phaff, Martin W. Miller, and Emil M. Mrak, *The Life of Yeasts: Second Revised and Enlarged Edition* (Cambridge: Harvard University Press, 1966), 1.

saw individual yeast cells: Ibid., 3.

21 *he worked at a big tax firm:* Dressler and Porter, *Discovering Enzymes,* 27.

Lavoisier ignored it: James A. Barnett and Linda Barnett, *Yeast Research: A Historical Overview* (London: ASM Press, 2011), 2.

out how fermentation worked: Ibid., 2.

in 1820s *pounds:* Ibid., 2.

physiologist named Theodor Schwann: Ibid., 6.

22 *left fruit juices alone:* Ibid., 7–8.

study in actual laboratories: Dressler and Porter, *Discovering Enzymes,* 40.

23 *"without needing a kidney":* Chemical Heritage Foundation, "Justus von Liebig and Friedrich Wöhler," http://www.chemheritage.org/discover/online-resources/chemistry-in-history/themes/molecular-synthesis-structure-and-bonding/liebig-and-wohler.aspx.

rearranged itself into ethanol: Dressler and Porter, *Discovering Enzymes,* 30.

things that weren't really there: Ibid., 31.

24 *another kind of fermentation:* Patrice Debré, *Louis Pasteur,* trans. Elborg Forster (Baltimore: Johns Hopkins University Press, 1994), 89.

he himself never drank much: Gerald L. Geison, *The Private Science of Louis Pasteur* (Princeton: Princeton University Press, 1995), 46.

"crystalline asymmetry, optical activity, and life": Ibid., 101.

25 *they were dirty or contaminated:* Brendon Barnett, "Fermentation," Pasteur Brewing, last modified December 29, 2011, http://www.pasteurbrewing.com/louis-pasteur-fermenation.html.

This was proof: Dressler and Porter, *Discovering Enzymes,* 33.

26 *proteins that accelerate biological processes:* Ibid., 38.

Pasteur kept up his end: Ibid., 35.

it was fermenting: Ibid., 48.

a whole bunch of purified enzymes: Ibid., 49.

27 *responsible for a given disease:* Barnett and Barnett, *Yeast Research,* 29.

The brewery switched: Ibid., 29.

designated it S. carlsbergensis: Ibid., 36.

every little detail of a fungus's looks: Charles Bamforth, *Scientific Principles of Malting and Brewing* (St. Paul, MN: American Society of Brewing Chemists, 2006), 36.

Sticky yeasts are harder: Sebastiaan Van Mulders et al., "The Genetics Behind Yeast Flocculation: A Brewer's Perspective," in *Yeast Flocculation, Vi-*

tality, and Viability: Proceedings of the 2nd International Brewers Symposium, ed. R. Alex Speers (St. Paul, MN: Master Brewers Association of the Americas, 2012), 36.

28 *stick to bubbles of carbon dioxide:* R. Alex Speers, "A Review of Yeast Flocculation," in *Yeast Flocculation, Vitality, and Viability: Proceedings of the 2nd International Brewers Symposium* (St. Paul, MN: Master Brewers Association of the Americas, 2012), 3.

flocculators don't floc anymore: Ibid., 5.

wild strains of yeast tend not to: Chris White and Jamil Zainasheff, *Yeast: The Practical Guide to Beer Fermentation* (Boulder, CO: Brewers Publications, 2010), 27.

29 *wolves turned into dogs:* Evan Ratliff, "Taming the Wild," *National Geographic,* March 2011, http://ngm.nationalgeographic.com/2011/03/taming-wild-animals/ratliff-text.

30 *the dogs expected the humans:* Jason G. Goldman, "Dogs, but Not Wolves, Use Humans as Tools," *The Thoughtful Animal* (blog), *Scientific American,* April 30, 2012, http://blogs.scientificamerican.com/thoughtful-animal/2012/04/30/dogs-but-not-wolves-use-humans-as-tools/.

eighty-one altogether: Justin C. Fay and Joseph A. Benavides, "Evidence for Domesticated and Wild Populations of *Saccharomyces cerevisiae,*" *PLoS Genetics* 1 (2005): 66–71.

32 *involving sugar metabolism:* Diego Libkind et al., "Microbe Domestication and the Identification of the Wild Genetic Stock of Lager-brewing Yeast," *Proceedings of the National Academy of Sciences* 108, no. 34 (August 22, 2011).

33 *Bacardi planned to start over:* W. Blake Gray, "Bacardi, and Its Yeast, Await a Return to Cuba," *Los Angeles Times,* October 6, 2011, http://latimes.com/features/food/la-fo-bacardi-20111006,0,1042.story.

a new roof over the old: Jacques De Keersmaecker, "The Mystery of Lambic Beer," *Scientific American* (August 1996), http://lambicandwildale.com/the-mystery-of-lambic-beer/.

34 *"the high points of my life":* Hiroichi Akiyama, *Saké: The Essence of 2000 Years of Japanese Wisdom Gained from Brewing Alcoholic Beverages from Rice,* trans. Inoue Takashi (Tokyo: Brewing Society of Japan. 2010), 95.

2. SUGAR

36 *what's now Kanazawa:* Joan Bennett, Presentation at the 2012 American Society for Microbiology Meeting (San Francisco: June 17, 2012).

he could speak Dutch: K. K. Kawakami, *Jokichi Takamine: A Record of His American Achievements* (New York: William Edwin Rudge, 1928), 1.

owned a sake brewery: Bennett, ASM presentation.

government-sponsored trip to Scotland: Kawakami, *Takamine,* 8.

all about enzymes or brewing: Akiyama, *Saké,* 115.

R. W. Atkinson's 1881 book: R. W. Atkinson, *The Chemistry of Sake Brewing* (Tokyo: Tokyo University, 1881).

37 *that doesn't mean he read it:* Akiyama, *Saké,* 115.

Korschelt, author of the 1878 article: William Shurtleff and Akiko Aoyagi, *History of Koji — Grains and/or Soybeans Enrobed with a Mold Culture (300 BCE to 2012): Extensively Annotated Bibliography and Sourcebook* (Lafayette, CA: Soyinfo Center, 2012), http://www.soyinfocenter.com/pdf/154/Koji.pdf, 6.

didn't teach Takamine: Akiyama, *Saké,* 115.

the key to sake, soy sauce: Katsuhiko Kitamoto, "Molecular Biology of the Koji Molds," *Advances in Applied Microbiology* 51 (January 2002): Table I.

characterized by a serious allergic reaction: National Library of Medicine, "Aspergillosis," last modified May 19, 2013, http://www.ncbi.nlm.nih.gov/pubmedhealth/PMH0002302/.

both toxic and carcinogenic: KeShun Liu, "Chemical Composition of Distillers Grains, a Review," *Journal of Agricultural and Food Chemistry* 59 (March 9, 2011): 1521.

as early as 300 BC: Shurtleff and Aoyagi, *Koji,* 5.

in Japanese records in 725: Ibid., 6.

making and selling koji: Ibid., 36.

businesses called moyashi: Masayuki Machida, Osamu Yamada, and Katsuya Gomi, "Genomics of *Aspergillus oryzae*: Learning from the History of Koji Mold and Exploration of Its Future," *DNA Research* 15 (August 2008): 174.

38 *Form glucose molecules into sheets:* Charles Bamforth, *Scientific Principles,* 23–24. Peter W. Atkins, *Molecules* (New York: Scientific American Library, 1987), 105, 102.

most common organic molecule: Ibid., 95.

39 *rabbits eat the poop:* Ibid., 102.

commonly known as starch: James S. Hough, *The Biotechnology of Malting and Brewing* (Cambridge: Cambridge University Press, 1985), 28.

Ferment and distill molasses: John E. Murtagh, "Feedstocks, Fermentation and Distillation for Production of Heavy and Light Rums," in *The Alcohol Textbook: A Reference for the Beverage, Fuel and Industrial Alcohol Industries,* ed. K. A. Jacques, T. P. Lyons, and D. R. Kelsall (Nottingham: Nottingham University Press, 1999), 243–55.

more fermentable lactose than milk: Sandor Katz, *The Art of Fermentation* (White River Junction, VT: Chelsea Green Publishing, 2012), 197.

In Sudan, they use: Ibid., 198.

Those substrates have a needle-burying: Hornsey, *Beer and Brewing,* 7.

40 *tastes like eggs and crocodile fat:* Roger G. Noll, "The Wines of West Africa: History, Technology and Tasting Notes," *Journal of Wine Economics* 3 (2008): 91–92.

50 to 250 gallons of glucose: Ian S. Hornsey, *Alcohol and Its Role in the Evolution of Human Society* (Cambridge: Royal Society of Chemistry, 2012), 467.

a viscous sludge called a biofilm: Katz, *Fermentation,* 90.
inulin breaks into fructose molecules: Hornsey, *Alcohol and Its Role,* 467.
Ferment and distill goop: Geoff H. Palmer, "Beverages: Distilled," *Encyclopedia of Grain Science,* ed. Colin Wrigley, Harold Corke, and Charles E. Walker (San Diego: Academic Press, 2004), 101.
that barley would thrive: Mark A. Harrison, "Beer/Brewing," in *Encyclopedia of Microbiology,* ed. Moselio Schaechter (San Diego: Academic Press, 2009), 24.

41 *They'll ferment on the vine:* Ian S. Hornsey, *The Chemistry and Biology of Winemaking* (Cambridge: Royal Society of Chemistry, 2007), 2.

42 *the same species of grape:* Hornsey, *Chemistry and Biology of Wine,* 68.
now the country Georgia: A. M. Negrul, "Method and Apparatus for Harvesting Grapes," US Patent 3,564,827 (Washington, DC: US Patent and Trade Office, 1971).
rise to western European grape varietals: R. Arroyo-García et al., "Multiple Origins of Cultivated Grapevine (*Vitis vinifera* L. Ssp. *Sativa*) Based on Chloroplast DNA Polymorphisms," *Molecular Ecology* 15 (October 2006): 3708.
extended vines called lianes: Hornsey, *Chemistry and Biology of Winemaking,* 75.
a yeast-friendly mix: Jaime Goode, *The Science of Wine* (Berkeley: University of California Press, 2005), 21.

43 *float around in the pulp:* Steven T. Lund and Joerg Bohlmann, "The Molecular Basis for Wine Grape Quality — A Volatile Subject," *Science* 311 (February 10, 2006), 804.
wild grapes are interesting: Hornsey, *Chemistry and Biology of Winemaking,* 79.
sweetest and most attractive to birds: Goode, *Science of Wine,* 21.

44 *late-ripening Petit Verdot:* Olivier Gergaud and Victor Ginsburgh, "Natural Endowments, Production Technologies and the Quality of Wines in Bordeaux. Does Terroir Matter?" *Journal of Wine Economics* 5 (2010): 15.

45 *an apartment in the French Quarter:* Joan W. Bennett, "Adrenalin and Cherry Trees," *Modern Drug Discovery* 4, no. 12 (December 2001), http://pubs.acs.org/subscribe/archive/mdd/v04/i12/html/12timeline.html, 47.
Tokyo Artificial Fertilizer Company: Kawakami, *Takamine,* 17.
They moved to Tokyo: Ibid., 18.
not on rice but on wheat bran: Jokichi Takamine, "Enzymes of Aspergillus Oryzae and the Application of Its Amyloclastic Enzyme to the Fermentation Industry," *Industrial & Engineering Chemistry* 6, no. 12 (1914): 825.
isolated starch-breaking enzymes: Bennett, ASM presentation.

46 *produced more sugar than malting:* Akiyama, *Saké,* 118.
the biggest distilling conglomerate in the United States: Kawakami, *Takamine,* 22.
"Her mother-in-law": Bennett, ASM presentation.

"She really was quite miserable": Ibid.

47 *a fifth of whatever profits:* Shurtleff and Aoyagi, *History of Koji,* 95.

the megabrand Johnnie Walker: "The Singleton Distilleries: Glen Ord," *Whisky Advocate* (Spring 2013): 97.

"A whisky that is pleasant enough": "Singleton of Glen Ord," *Whisky News* (blog), http://malthead.blogspot.com/2006/12/singleton-of-glen-ord_10.html.

48 *Corn needs too much heat:* Hornsey, *History of Beer,* 15. Hough, *Biotechnology of Malting and Brewing,* 8.

50 *a wall called a scutellum:* Ibid., 13.

encased in a hard cellulose: Bamforth, *Scientific Principles,* 45, 23.

52 Oryza sativa: Akiyama, *Saké,* 48.

inhibit the growth of yeast: Ibid., 50.

53 *chewing manioc or cornmeal:* Hornsey, *History of Beer,* 27–28.

1876, to be specific: Machida, "Genomics of *Aspergillus oryzae,*" 175.

to have its genes sequenced: Goffeau et al., "6000 Genes," 546.

koji didn't get its turn: Masayuki Machida et al., "Genome Sequencing and Analysis of *Aspergillus oryzae,*" *Nature* 438 (December 22, 2005): 1157.

20 million years old: Ibid., 1160.

three distinct alpha-amylases: Ibid., 1159.

Yet A. oryzae *seems perfectly safe:* John G. Gibbons et al., "The Evolutionary Imprint of Domestication on Genome Variation and Function of the Filamentous Fungus *Aspergillus oryzae,*" *Current Biology* 22 (2012): 1.

54 *those frozen regions:* Ibid., 2.

55 *the malt manufacturers campaigned*: Kawakami, *Takamine,* 29.

56 *firemen found there was no pressure:* Shurtleff and Aoyagi, *Koji,* 95.

whisky called Bonzai: Bennett, "Adrenalin and Cherry Trees," 48.

They had to ask their families: Bennett, ASM presentation.

the Whiskey Trust broke it off: Kawakami, *Takamine,* 30.

"the Alka-Seltzer of the 1890s": Bennett, ASM presentation.

pharma company Parke-Davis: Kawakami, *Takamine,* 36.

57 *Judge Learned Hand ruled:* Bennett, "Adrenalin and Cherry Trees," 51.

decorated the first two floors: Kawakami, *Takamine,* 65–67.

the Tidal Basin in Washington, D.C.: Bennett, "Adrenalin and Cherry Trees," 51.

3. FERMENTATION

59 *human-made fermented beverage on earth:* Patrick E. McGovern et al., "Fermented Beverages of Pre- and Proto-historic China," *Proceedings of the National Academy of Sciences* 101, no. 51 (December 2004): 17593.

an evanescent molecule: McGovern et al., "Fermented Beverages," 17597.

60 *all they could taste was water:* Patrick E. McGovern, *Ancient Wine: The*

Search for the Origins of Viniculture (Princeton: Princeton University Press, 2007), 52.

other molecules unique to fermented beverages: Ibid., 62. Rudolph H. Michel, Patrick E. McGovern, and Virginia R. Badler, "The First Wine & Beer: Chemical Detection of Ancient Fermented Beverages," *Analytical Chemistry* 65 (April 1993): 408A–13A.

wine-like red residue: McGovern, *Ancient Wine,* 40.

A suite of chemical traces: Michel, McGovern, and Badler, "The First Wine & Beer," 408A.

possibly domesticated rice: McGovern et al., "Fermented Beverages," 17593.

61 *Oliver Dietrich:* Oliver Dietrich et al., "The Role of Cult and Feasting in the Emergence of Neolithic Communities. New Evidence from Gobekli Tepe, South-eastern Turkey," *Antiquity* 86 (2012): 687.

62 *Its customers are mostly home brewers:* White Labs, "About White Labs," http://www.whitelabs.com/about_us.html.

WLP802 Czech Budejovice Lager: White Labs, "Professional Yeast Bank," http://www.whitelabs.com/beer/craft_strains.html.

63 *Different yeasts make better-tasting stuff:* Barbara Dunn and Gavin Sherlock, "Reconstruction of the genome origins and evolution of the hybrid lager yeast *Saccharomyces pastorianus,*" *Genome Research* 18 (2008): 1610.

the equally iconic Burton upon Trent: Bamforth, *Scientific Principles,* 80.

porter to lagers like pilsner and bock: Bamforth, *Scientific Principles,* 10.

64 *a whopping 152.19 parts per million:* White and Zainasheff, *Yeast,* 96.

65 *the stuff that keeps the lights on:* Rosaura Rodicio and Jürgen J. Heinisch, "Sugar Metabolism by Saccharomyces and Non-*Saccharomyces* Yeasts," in *Biology of Microorganisms on Grapes, in Must, and in Wine,* ed. H. König (Berlin: Springer-Verlag, 2009), 123.

the same stuff in pickles: Garry Menz et al., "Isolation, Identification, and Characterisation of Beer-spoilage Lactic Acid Bacteria from Microbrewed Beer from Victoria, Australia," *Journal of the Institute of Brewing* 116, no. 1 (2010): 14.

making ethanol and ATP: Bamforth, *Scientific Principles,* 105.

then the ethanol diffuses: Rodicio and Heinisch, "Sugar Metabolism," Figure 6.4 caption.

66 *Acetaldehyde appears and disappears:* Sebastian Meier et al., "Metabolic Pathway Visualization in Living Yeast by DNP-NMR," *Molecular bioSystems* 7, no. 10 (October 2011): 2835.

67 *they turn ethanol into acetaldehyde:* J. Michael Thomson et al., "Resurrecting Ancestral Alcohol Dehydrogenases from Yeast," *Nature Genetics* 37, no. 6 (June 2005): 630.

69 *living cells were transforming organic compounds:* Dressler and Porter, *Discovering Enzymes,* 34–35. Gerald L. Geison, *Louis Pasteur,* 107.

come from pigments: Hornsey, *Chemistry and Biology of Winemaking,* 79.
barleys intended for ales: Bamforth, *Scientific Principles,* 17.
"very different flavor profiles": Isak S. Pretorius, Christopher D. Curtin, and Paul J. Chambers, "The Winemaker's Bug: From Ancient Wisdom to Opening New Vistas with Frontier Yeast Science," *Bioengineered Bugs* 3, no. 3 (2012): 150.

70 *variations up to a thousandfold:* Chandra L. Richter et al., "Comparative Metabolic Footprinting of a Large Number of Commercial Wine Yeast Strains in Chardonnay Fermentations," *FEMS Yeast Research* 13, no. 4 (June 2013): 394.

71 *keep the same yeast strain consistent:* Clay Risen, "Whiskey Myth No. 2," *Mash Notes* (blog), last modified July 27, 2012, http://clayrisen.com/?p=126.
build your distillery right next door: Robert Piggot, "Rum: Fermentation and Distillation," in *The Alcohol Textbook,* 5th ed., ed. W. M. Ingeldew et al. (Nottingham: Nottingham University Press, 2009), 476.

72 *Puerto Rican government built him a lab:* Rafael Arroyo, *Studies on Rum,* Research Bulletin no. 5 (Rio Piedras: University of Puerto Rico Agricultural Experimental Station, December 1945): 3
collected and isolated a mold: Ibid., 94.

73 *have entirely other colonizers:* Nicholas A. Bokulich et al., "Microbial Biogeography of Wine Grapes Is Conditioned by Cultivar, Vintage, and Climate," *Proceedings of the National Academies of Science* (published online ahead of print November 25, 2013): 2.
Carbon dioxide has its own flavor: Jayaram Chandrashekar et al., "The Taste of Carbonation," *Science* 326, no. 5951 (October 16, 2009): 443.

74 *inject it back into the beer:* White and Zainasheff, *Yeast,* 115.
a conical jet into the headspace: Gérard Liger-Belair, *Uncorked: The Science of Champagne* (Princeton: Princeton University Press, 2004), 88.
the wine's perceived aroma: Gérard Liger-Belair et al., "Unraveling Different Chemical Fingerprints Between a Champagne Wine and Its Aerosols," *Proceedings of the National Academies of Science* 106, no. 39 (2009): 16548.
5 grams of CO_2 per liter: Charles W. Bamforth, *Foam* (St. Paul, MN: American Society of Brewing Chemists), 8.
champagne is pressurized: Liger-Belair, *Uncorked,* 37.

75 *propel a popped champagne cork:* Gérard Liger-Belair, Guillaume Polidori, and Philippe Jeandet, "Recent Advances in the Science of Champagne Bubbles," *Chemical Society Reviews* 37, no. 11 (November 2008): 2493.
liquid molecules stick together: Liger-Belair, *Uncorked,* 40.
University of Reims in France: Richard Ingham, "Champagne Physicist Reveals the Secrets of Bubbly," *Phys.org,* last modified September 18, 2012, http://phys.org/news/2012-09-champagne-physicist-reveals-secrets.html.
3,000 frames per second: Liger-Belair, *Uncorked,* 41–42.
bubbles to come from scratches: Ibid., 41–42.

glassmakers laser-etch tiny nucleation sites: Bamforth, *Foam,* 9.
the inside of his champagne flutes: Liger-Belair, *Uncorked,* 44.
they became "bubble guns": Ibid., 51.
thirty bubbles per second: Ibid., 55.
loses CO_2 faster than a tall, narrow flute does: Ingham, "Champagne Physicist."
hit the surface and pop: Liger-Belair et al., "Chemical Fingerprints," 16545.
77 *Bamforth and a colleague discovered:* Ibid., 16.
points outward into the liquid: James J. Hackbarth, "Multivariate Analyses of Beer Foam Stand," *Journal of the Institute of Brewing* 112, no. 1 (2006): 17.
Foam is a collaboration among bubbles: Bamforth, *Foam.* 10.
gravity pulls the beer back down: Hackbarth, "Multivariate Analyses," 18.
78 *McGovern found in a 2,700-year-old tomb:* Dogfish Head Brewing, "Midas Touch," http://www.dogfish.com/brews-spirits/the-brews/year-round-brews/midas-touch.htm.
wild yeast from an Egyptian date farm: Abigail Tucker, "The Beer Archaeologist," *Smithsonian,* July–August 2011, http://www.smithsonianmag.com/history-archaeology/The-Beer-Archaeologist.html?c=y&story=fullstory.
79 *"experimental archaeology":* Tucker, "Beer Archaeologist."

4. DISTILLATION

82 *Kininvie used the same barley:* Jim Murray, "Tomorrow's Malt," *Whisky* 1 (1999): 56.
84 *Primo Levi:* Bruce Moran, *Distilling Knowledge: Alchemy, Chemistry, and the Scientific Revolution* (Cambridge: Harvard University Press, 2005), 12.
still-like devices in China around 3000 BC: Robert Piggot, "Beverage Alcohol Distillation," in *The Alcohol Textbook,* 5th ed., ed. W. M. Ingeldew et al. (Nottingham: Nottingham University Press, 2009), 431.
85 *a high-enough alcohol concentration to catch fire:* H. T. Huang, *Science and Civilisation in China,* vol. 6, pt. 5, *Biology and Biological Technology: Fermentations and Food* (Cambridge: Cambridge University Press, 2000), 206.
written almost 600 years later: C. Anne Wilson, *Water of Life: A History of Wine-Distilling and Spirits 500 BC to AD 2000* (Devon, UK: Prospect Books, 2006), 38.
plausibly in Roman times: Piggot, "Beverage Alcohol Distillation," 431.
"India appears on present evidence": F. R. Allchin, "India: The Ancient Home of Distillation?" *Man* 14, no. 1 (1979): 59–63.
called Maria the Jewess: Raphael Patai, *The Jewish Alchemists* (Princeton: Princeton University Press, 1994), 3.
86 *first museum on earth:* Justin Pollard and Howard Reid, *The Rise and Fall of Alexandria: Birthplace of the Modern Mind* (New York: Viking, 2006), 59.
stored in a golden box: Ibid., 47.

prevailing winds to keep cool: Ibid., 24.
the smallest indivisible pieces of matter: Ibid., xvii.
"formed the library of Alexandria": Ibid., xvi.
burning 400,000 book scrolls: Ibid., 165.

87 *a home to Alexandrian scholarship:* Judith McKenzie, *The Architecture of Alexandria and Egypt 300 BC–AD 700* (New Haven: Yale University Press, 2007), 54–56.
a giant statue of the god Serapis: Pollard and Reid, *Rise and Fall of Alexandria,* 203.
His clockwork designs: Ibid., 179.
a trumpet-playing robot: Ibid., 179–85.
a device called an aeolipile: Ibid., 188.
It was a city of engineers: Ibid., 178

88 *the historian Raphael Patai:* Patai, *Jewish Alchemists,* 60.
women chemists developed various cosmetics: Elizabeth H. Oakes, *Encyclopedia of World Scientists* (New York: Infobase Learning, 2007), 485.
the city had multiple synagogues: Theodore Vrettos, *Alexandria: City of the Western Mind* (New York: The Free Press, 2001), 163.
"you are not of the race of Abraham": Patai, *Jewish Alchemists,* 69–70.
convince early Christians she was a prophet: Ibid., 66.

89 *to describe a specific kind of still:* R. J. Forbes, *Short History of the Art of Distillation* (Leiden, the Netherlands: E. J. Brill, 1948), 23.

90 *He burned the Egyptian chemistry books, too:* McKenzie, *Architecture of Alexandria and Egypt,* 209.
turned the Serapeum into a church: Sarah Zielinski, "Hypatia, Ancient Alexandria's Great Female Scholar," *Smithsonian.com,* last modified March 15, 2010, http://www.smithsonianmag.com/history-archaeology/Hypatia-Ancient-Alexandrias-Great-Female-Scholar.html.
The amphorae that carried it: Bouby et al., "Bioarchaeological Insights," e63195.

91 *"bread wine":* Robert Piggot, "Vodka, Gin and Liqueurs," in *The Alcohol Textbook,* 5th ed., ed. W. M. Ingeldew et al. (Nottingham: Nottingham University Press, 2009), 465.
Salernus mentions the use of a still: Forbes, *History of Distillation,* 57.
De secretis mulierum: Ibid., 58.
"aguardentes" or "aguardientes": Wilson, *Water of Life,* 102.
out the essences of plants: Ibid., 115.
food historian C. Anne Wilson: Ibid., 116.

92 *That's the boiling point:* Richard J. Panek and Armond R. Boucher, "Continuous Distillation," in *The Science and Technology of Whiskies,* ed. J. R. Piggott, R. Sharp, and R. E. B. Duncan (London: Longman Group, 1989), 151.
It's called the azeotropic limit: Ibid., 152.
mix gunpowder into the rum: D. Nicol, "Batch Distillation," in *The Science*

and Technology of Whiskies, ed. J. R. Piggott, R. Sharp, and R. E. B. Duncan (London: Longman Group, 1989), 132.

93 *Unless you're making vodka:* Panek and Boucher, "Continuous Distillation," 152.

corrupted to "whisky": E. C., "On the Antiquity of Brewing and Distillation in Ireland," *Ulster Journal of Archaeology* 7 (1859): 34.

Liquor had spread across the world: Wilson, *Water of Life,* 147–51.

pot stills chained together: J. E. Murtagh, "Feedstocks, Fermentation and Distillation," 245.

aquavit comes off a continuous still: Palmer, "Beverages: Distilled," 97.

94 *A river of fire:* Charles K. Cowdery, *Bourbon, Straight: The Uncut and Unfiltered Story of American Whiskey* (Chicago: Made and Bottled in Kentucky, 2004), 87.

95 *continuous distillation using a column:* Forbes, *History of Distillation,* 298.

fractional distillation: Marco Lagi and R. S. Chase, "Distillation: Integration of a Historical Perspective," *Australian Journal of Education in Chemistry* 70 (2009): 7.

technology used to crack crude oil: Ibid., 10.

Steam rushes upward: Wilson, *Water of Life,* 256–57.

have a metallic flavor: Nicol, "Batch Distillation," 137.

The main still in a big bourbon distillery: Panek and Boucher, "Continuous Distillation," 154–56.

96 *yet another subsidiary still:* Cowdery, *Bourbon, Straight,* 12.

more or less volatile: Panek and Boucher, "Continuous Distillation," 154.

Other than the shape of the still: Nicol, "Batch Distillation," 134.

98 *You end up with hydrogen sulfide:* Barry Harrison et al., "The Impact of Copper in Different Parts of Malt Whisky Pot Stills on New Make Spirit Composition and Aroma," *Journal of the Institute of Brewing* 117, no. 1 (2011): 106.

other sulfur-containing compounds: Ibid., 109.

99 *maintenance checks every year on the stills:* Matthew Rowley, "Replacing That Worn Out Still — Every Ding and Dent?" *Rowley's Whiskey Forge* (blog), last modified January 17, 2013, http://matthew-rowley.blogspot.com/2013/01/replacing-that-worn-out-still-every.html.

100 *twenty years before, there were only four :* Michael Kintslick, *The U.S. Craft Distilling Market: 2011 and Beyond* (New York: Coppersea Distillery, 2011), 1.

5. AGING

109 *storing wine in wooden barrels:* McGovern, *Ancient Wine,* 167.

"date palms do not grow there": Ibid., 167–68.

Booze chemistry changes over time: Hornsey, *Chemistry and Biology of Winemaking,* 39–40.

110 *a ceramic jar called a* dolium: Alexander Conison, "The Organization

of Rome's Wine Trade," PhD diss. (University of Michigan: 2012), 169–72, http://deepblue.lib.umich.edu/bitstream/handle/2027.42/91455/conison_1.pdf?sequence=1.

Maybe they used them to store wine: Hornsey, *Chemistry and Biology of Winemaking,* 40.

bend wood into a curved shape: McGovern, *Ancient Wine,* 260–62.

111 *But "Old Bourbon" referred:* Cowdery, *Bourbon, Straight,* 27.

Cowdery's research says: Ibid., 31.

"probably were evolutionary": Ibid., 32–33.

113 *The barrels are all oak:* J. R. Mosedale and Jean-Louis Puech, "Barrels: Wines, Spirits, and Other Beverages," in *Encyclopedia of Food Sciences and Nutrition,* ed. Benjamin Caballero, Luiz Trugo, and Paul Finglass (San Diego: Academic Press, 2003), 393.

wood seasoned outside: Independent Stave Company, *International Barrel Symposium: Research Results and Highlights from the 5th, 6th, and 7th Symposiums* (Lebanon, MO: 2008), 8.

It's a ring of wood staves: Ibid., 19.

114 *soluble in ethanol or water:* Mosedale and Puech, "Barrels," 394.

turns into phenols: Ibid., 395.

big and meaty flavor: Ibid., 395.

oak tannins: Independent Stave Company, 8.

thermoplastic properties of the wood: Mosedale and Puech, "Barrels," 398.

119 *half of them are vanillin:* J. R. Mosedale and Jean-Louis Puech, "Wood Maturation of Distilled Beverages," *Trends in Food Science & Technology* 9, no. 3 (March 1998): 96–97.

the same flavors as browned meat: Mosedale and Puech, "Barrels," 400–401.

120 *Incoming oxygen means oxidation:* Mosedale and Puech, "Wood Maturation," 97.

beer is full of lipids: B. Vanderhaegen et al., "The Chemistry of Beer Aging—A Critical Review," *Food Chemistry* 95, no. 3 (April 2006): 358.

the aroma of the spirit: Melina Macatelli, John R. Piggott, and Alistair Paterson, "Structure of Ethanol-Water Systems and Its Consequences for Flavour," in *New Horizons: Energy, Environment, and Enlightenment: Proceedings of the Worldwide Distilled Spirits Conference,* ed. G. M. Walker and P. S. Hughes (Nottingham: Nottingham University Press, 2010), 236.

people call a drink "smooth": Mosedale and Puech, "Wood Maturation," 97.

122 *"The ever increasing demands of the trade":* Arroyo, *Studies on Rum,* 169.

flavor of fine French brandy: Hewson Clarke and John Dougall, *The Cabinet of Arts, or General Instructor in Arts, Science, Trade, Practical Machinery, the Means of Preserving Human Life, and Political Economy, Embracing a Variety of Important Subjects* (London: T. Finnersley, 1817), 722.

123 *"This will make malt liquor":* William T. Boothby, *Cocktail Boothby's Ameri-*

can Bar-Tender (San Francisco: H. S. Crocker, 1891; San Francisco: Anchor Distilling, 2008): 365.

The very same wording: Our Knowledge Box: Or, Old Secrets and New Discoveries (New York: Geo. Blackie and Co., 1875), http://www.gutenberg.org/files/43418/43418-h/43418-h.htm#SECRETS_OF_THE_LIQUOR_TRADE, unpaged.

spray the stuff into the barrel: J. M. Philp, "Cask Quality and Warehouse Conditions," in *The Science and Technology of Whiskies,* ed. J. R. Piggott, R. Sharp, and R. E. B. Duncan (London: Longman Group, 1989), 273–74.

the process was called tranchage: Mosedale and Puech, "Wood Maturation," 100.

electrolysis and ultraviolet light: Arroyo, *Studies on Rum,* 170.

artificial aging technologies in rum: J. Quesada Granados et al., "Application of Artificial Aging Techniques to Samples of Rum and Comparison with Traditionally Aged Rums by Analysis with Artificial Neural Nets," *Journal of Agricultural and Food Chemistry* 50, no. 6 (March 13, 2002): 1471.

124 *high heat and humidity in Taiwan:* Dominic Roskrow, "Is It the Age? Or the Mileage?" *Whisky Advocate* (Winter 2011): 77–80.

126 *The bourbon distiller Buffalo Trace:* Buffalo Trace Distillery, *Warehouse X* (blog), http://www.experimentalwarehouse.com.

6. SMELL AND TASTE

135 *"Since there are many wine writers":* Richard E. Quandt, "On Wine Bullshit: Some New Software?" *Journal of Wine Economics* 2, no. 2 (2007): 130.

"ties the odor to its physical source": Donald A. Wilson and Richard J. Stevenson, *Learning to Smell: Olfactory Perception from Neurobiology to Behavior* (Baltimore: Johns Hopkins University Press, 2006), 7.

136 *"sloppy," and "vivacious":* Adrienne Lehrer, *Wine & Conversation,* 2nd ed. (New York: Oxford University Press, 2009), 16.

137 *mix of every wavelength:* Tali Weiss et al., "Perceptual Convergence of Multi-Component Mixtures in Olfaction Implies an Olfactory White," *Proceedings of the National Academy of Sciences* 109, no. 49 (2012): 19959.

Andrew Livermore and David Laing: Andrew Livermore and David G. Laing, "The Influence of Chemical Complexity on the Perception of Multicomponent Odor Mixtures," *Perception & Psychophysics* 60, no. 4 (May 1998): 650.

138 *subjects could pick out the tastes:* K. Marshall et al., "The Capacity of Humans to Identify Components in Complex Odor-Taste Mixtures," *Chemical Senses* 31, no. 6 (July 2006): 543.

"The Color of Odors": Gil Morrot, Frédéric Brochet, and Denis Dubourdieu, "The Color of Odors," *Brain and Language* 79, no. 2 (2001): 309–20.

139 *The four-level training to be a master sommelier:* Court of Master Sommeliers, "Courses & Schedules," http://www.mastersommeliers.org/Pages.aspx/Master-Sommelier-Diploma-Exam.

141 *University of Padua in Italy and Macquarie University in Australia:* Gesualdo Zucco et al., "Labeling, Identification, and Recognition of Wine-relevant Odorants in Expert Sommeliers, Intermediates, and Untrained Wine Drinkers," *Perception* 40, no. 5 (2011): 598–607.

143 *receptors for both sweet and bitter:* Anna Scinska et al., "Bitter and Sweet Components of Ethanol Taste in Humans," *Drug and Alcohol Dependence* 60, no. 2 (August 1, 2000): 205.

polymodal nociceptors: Alexander A. Bachmanov et al., "Chemosensory Factors Influencing Alcohol Perception, Preferences, and Consumption," *Alcoholism: Clinical and Experimental Research* 27, no. 2 (February 2003): 227.

A little pain mixes into both: John Leffingwell, "Update No. 5: Olfaction," *Leffingwell Reports* 2 (May 2002): 3.

Flavor is a combination of taste and smell: Lehrer, *Wine & Conversation,* 7.

subjects spit out the tastant: Marcia Levin Pelchat and Fritz Blank, "A Scientific Approach to Flavours and Olfactory Memory," in *Food and the Memory: Proceedings of the Oxford Symposium on Food and Cookery,* ed. Harlan Walker (Devon, UK: Prospect Books, 2001), 187.

144 *Olfactory receptors thread back and forth:* Bettina Malnic, Daniela C. Gonzalez-Kristeller, and Luciana M. Gutiyama, "Odorant Receptors," in *The Neurobiology of Olfaction,* ed. Anna Menini (Boca Raton, FL: CRC Press, 2010), 183.

That's a lot of processing power: Ibid., 184.

Processing of smells in the brain: Donald A. Wilson and Robert L. Rennaker, "Cortical Activity Evoked by Odors," in *The Neurobiology of Olfaction,* ed. Anna Menini (Boca Raton, FL: CRC Press, 2010), 354.

145 *red wine taste better:* Ellena S. King, Randall L. Dunn, and Hildegarde Heymann, "The Influence of Alcohol on the Sensory Perception of Red Wines," *Food Quality and Preference* 28, no. 1 (April 2013): 243.

146 *Protons are one of the hallmarks:* Chandrashekar et al., "The Taste of Carbonation," 444.

148 *spiking a sweet drink:* Bachmanov et al., "Chemosensory Factors," 225.

super-tasters report lower levels of alcohol: Ibid., 228.

149 *each with a new schema:* Joseph N. Kaye, "Symbolic Olfactory Display," PhD diss. (Massachusetts Institute of Technology, 2001).

It didn't work: Wilson and Stevenson, *Learning to Smell,* 12–13.

"burnt" was guaiacol: Ibid., 13.

something of an evangelist: "A Wake for Morten Christian Meilgaard," *Flower Parties through the Ages* (blog), http://goodfelloweb.com/flowerparty/fp_2009/Morten_Meilgaard_1928-2009.htm.

151 *methanol and isopropanol:* Luke N. Rodda et al., "Alcohol Congener Analysis and the Source of Alcohol: A Review," *Forensic Science, Medicine, and Pathology* 9, no. 2 (June 2013): 199.

152 *hydrogen bond strength:* Hu et al., "Structurability," 7398.

help guide the industry today: Arroyo, *Studies on Rum,* 3.

157 *request wine for science:* A. L. Robinson et al., "Influence of Geographic Origin on the Sensory Characteristics and Wine Composition of *Vitis vinifera* Cv. Cabernet Sauvignon Wines from Australia," *American Journal of Enology and Viticulture* 63, no. 4 (June 25, 2012): 467–76.

159 *glass shape has a limited effect:* J. F. Delwiche and Marcia Levin Pelchat, "Influence of Glass Shape on Wine Aroma," *Journal of Sensory Studies* 17, no. 2002 (2002): 28. T. Hummel et al., "Effects of the Form of Glasses on the Perception of Wine Flavors: A Study in Untrained Subjects," *Appetite* 41, no. 2 (October 2003): 201.

pint glasses with curved sides: Angela S. Attwood et al., "Glass Shape Influences Consumption Rate for Alcoholic Beverages," *PLoS ONE* 7, no. 8 (January 2012): e43007.

7. BODY AND BRAIN

160 *and a serial marryer:* Dennis Hevesi, "G. A. Marlatt, Advocate of Shift in Treating Addicts, Dies at 69," *New York Times,* March 21, 2011, http://www.nytimes.com/2011/03/22/us/22marlatt.html?_r=0.

"Loss of Control Drinking in Alcoholics: An Experimental Analogue": G. Alan Marlatt, Barbara Demming, and John B. Reid, "Loss of Control Drinking in Alcoholics: An Experimental Analogue," *Journal of Abnormal Psychology* 81, no. 3 (1973): 233–41.

161 *isolating the effects of alcohol on people:* G. Alan Marlatt, "This Week's Citation Classic," *ISI Current Contents: Social and Behavioral Sciences* 18 (May 6, 1985): 18.

because combined with a mixer: Marlatt, Demming, and Reid, "Loss of Control Drinking," 234.

"We spent many a pleasant evening experimenting": Marlatt, "Citation Classic," 18.

162 *Behavioral Alcohol Research Laboratory:* Daniel Wood, "Bar Lab Challenges the Alcohol Mystique," *Chicago Tribune,* February 24, 1991, http://articles.chicagotribune.com/1991-02-24/features/9101170848_1_addictive-behaviors-research-center-alcohol-free-alcoholism-and-alcohol-abuse.

the expect-ethanol/get-placebo group: Stephen A. Maisto, Gerard J. Connors, and Paul R. Sachs, "Expectation as a Mediator in Alcohol Intoxication: A Reference Level Model," *Cognitive Therapy and Research* 5, no. 1 (March 1981): 7.

their own internal state of intoxication: Ibid., 11–12.

164 *The volunteer identified only as* Ius: Walter R. Miles, *Alcohol and Human Efficiency* (Washington, DC: Carnegie Institute of Washington, 1924), 113.
"Effect of Dilute Alcohol Given by Rectal Injection During Sleep": Ibid., 111.

165 *"did not waken the subject"*: Ibid., 111.

166 *the stomach and upper intestine:* Nischita K. Reddy, Ashwani Singal, and Don W. Powell, "Alcohol-Related Diarrhea," in *Diarrhea: Diagnostic and Therapeutic Advances,* ed. Stefano Guandalini and Haleh Vaziri (New York: Springer, 2011), 381.
"Proposed Tentative Program for an Investigation of the Physiological Effects of Alcohol": Francis G. Benedict and Raymond Dodge, *Psychological Effects of Alcohol; an experimental investigation of the effects of moderate doses of ethyl alcohol on a related group of neuro-muscular processes in man* (Washington, DC: Carnegie Institute of Washington, 1915), 266.
Benedict soon turned a room: Ibid., 30.

167 *An L-shaped wooden device:* Ibid., 38.
"a satisfactory adhesive medium": Ibid., 58.
paper eyelashes on people: Ibid., 59.
Nothing worked: Ibid., 22.
gadget-heavy experiment: E. R. Hilgard, *Walter Richard Miles, 1885–1978* (Washington, DC: National Academy of Sciences, 1985), http://www.nasonline.org/publications/biographical-memoirs/memoir-pdfs/miles-walter.pdf.
"this pharmacodynamic substance": Miles, *Alcohol and Human Efficiency,* 272.

168 *in your bloodstream as ethanol:* Harold E. Himwich, "The Physiology of Alcohol," *Journal of the American Medical Association* 1446, no. 7 (1957): 545.
at least one cancer-causing chemical: Samir Zakhari, "Overview: How Is Alcohol Metabolized by the Body?" *Alcohol Research & Health* 29, no. 4 (January 2006): 246.

169 *ethanol nearly doubles that:* Hornsey, *Beer and Brewing,* 6. Antonow and McClain, "Nutrition and Alcoholism," 81.
Alcoholics get up to 50 percent: Robert Dudley, "Evolutionary Origins of Human Alcoholism in Primate Frugivory," *Quarterly Review of Biology* 75, no. 1 (2000): 6.

170 *some nasty tummy symptoms:* Mimy Y. Eng, Susan E. Luczak, and Tamara L. Wall, "ALDH2, ADH1B, and ADH1C Genotypes in Asians: A Literature Review," *Alcohol Research & Health* 30, no. 1 (2007): 22–27.
higher rates of esophageal cancer: Benjamin Taylor and Jürgen Rehm, "Moderate Alcohol Consumption and Diseases of the Gastrointestinal System: A Review of Pathophysiological Processes," in *Alcohol and the Gastrointestinal Tract,* ed. Manfred Singer and David Brenner (Basel: Karger Publishers, 2006), 30.

you also throw up: Mercè Correa et al., "Piecing Together the Puzzle of Acetaldehyde as a Neuroactive Agent," *Neuroscience and Biobehavioral Reviews* 36, no. 1 (January 2012): 409.

ethanol can make you have to pee: Murray Epstein, "Alcohol's Impact on Kidney Function," *Alcohol Health & Research World* 21, no. 1 (1997): 85.

the body's overall concentration of electrolytes: Ibid., 85.

171 *a stretchy nylon cap:* Alan Gevins, Cynthia S. Chan, and Lita Sam-Vargas, "Towards Measuring Brain Function on Groups of People in the Real World," *PLoS ONE* 7, no. 9 (September 5, 2012): e44676.

174 *No one had ever tried it before:* Jennifer M. Mitchell et al., "Alcohol Consumption Induces Endogenous Opioid Release in the Human Orbitofrontal Cortex and Nucleus Accumbens," *Science Translational Medicine* 4, no. 116 (January 11, 2012): 116ra6.

177 *The song of the zebra finch:* Emma Young, "Silent Song," *New Scientist,* October 27, 2000, http://www.newscientist.com/article/dn110-silent-song.html.

The songs become disorganized: Sara Reardon, "Zebra Finches Sing Sloppily When Drunk," *New Scientist,* October 17, 2012, http://www.newscientist.com/article/dn22389-zebra-finches-sing-sloppily-when-drunk.html.

ethanol's numbing effects: Keith Johnson, David B. Pisoni, and Robert H. Bernacki, "Do Voice Recordings Reveal Whether a Person Is Intoxicated? A Case Study," *Phonetica* 47 (1990): 216.

replace Breathalyzers in remote testing: William Yang Wang, Fadi Biadsy, Andrew Rosenberg, and Julia Hirschberg, "Automatic Detection of Speaker State: Lexical, Prosodic, and Phonetic Approaches to Level-of-interest and Intoxication Classification," *Computer Speech & Language* 27 (April 2012), http://dx.doi.org/10.1016/j.csl.2012.03.004.

178 *they're blaming congeners:* Rodda, "Alcohol Congener Analysis," 203.

affect the brain and body differently: H. B. Haag et al., "Studies on the Acute Toxicity and Irritating Properties of the Congeners in Whisky," *Toxicology and Applied Pharmacology* 627, no. 6 (1959): 618.

179 *absinthism contributed to Vincent van Gogh's suicide:* Wilfred Niels Arnold, "Absinthe," *Scientific American,* June 1989, http://www.scientificamerican.com/article.cfm?id=absinthe-history.

the drink was illegal: Brian Ashcraft, "The Mystery of the Green Menace," *Wired,* November 2005. D. W. Lachenmeier et al., "Absinthe, Absinthism and Thujone — New Insight into the Spirit's Impact on Public Health," *Open Addiction Journal* 3 (2010): 33–34.

180 *a spontaneously formed emulsion:* N. L. Sitnikova et al., "Spontaneously Formed Trans-Anethol/Water/Alcohol Emulsions: Mechanism of Formation and Stability," *Langmuir* 21, no. 8 (2005): 7083.

But in the super-bourbon group: Henri Begleiter and Arthur Platz, "The Effects of Alcohol on the Central Nervous System in Humans," in *The Biology of Alcoholism,* ed. B. Kissin and Henri Begleiter (New York: Plenum Publishing Corporation, 1972), 325.

Some say it's red wine: Alessandro Panconesi, "Alcohol and Migraine: Trigger Factor, Consumption, Mechanisms: A Review," *Journal of Headache and Pain* 9, no. 1 (February 2008): 22.

difference among the various drinks: Ibid., 23.

181 *excitatory neurotransmitter norepinephrine:* Christine N. Jayarajah et al., "Analysis of Neuroactive Amines in Fermented Beverages Using a Portable Microchip Capillary Electrophoresis System," *Analytical Chemistry* 79, no. 21 (November 2007): 8162.

white wine nor red wine contain much tyramine: Panconesi, "Alcohol and Migraine," 23.

It blocks serotonin: Ibid., 24.

182 Drunken Comportment: A Social Explanation: Craig MacAndrew and Robert Edgerton, *Drunken Comportment* (Hawthorn, NY: Aldine, 1969; Clinton Corners, NY: Elliot Werner, 2003). Citations refer to the Elliot Werner edition.

widespread alcohol dependence problems: Ibid., 48.

183 *Taira villagers got happier:* MacAndrew and Edgerton, *Drunken Comportment,* 53–55.

"Even if, say, the brain physiologists": Ibid., 11–12.

186 How Much Is Too Much? The Effects of Social Drinking: Leonard Gross, *How Much Is Too Much? The Effects of Social Drinking* (New York: Magilla, 1983).

"I will not compromise": Ibid., 24–25.

8. HANGOVER

187 *veisalgia, from the Greek:* Jeffrey G. Wiese, Michael G. Shlipak, and Warren S. Browner, "The Alcohol Hangover," *Annals of Internal Medicine* 132, no. 11 (2000): 897–98.

188 *Elpenor is missing:* Ibid., 901.

$160 billion a year: Derek Thompson, "The Economic Cost of Hangovers," *The Atlantic,* July 5, 2013, http://www.theatlantic.com/business/archive/2013/07/the-economic-cost-of-hangovers/277546/.

189 *Hangover got 406 studies:* Joris C. Verster and Richard Stephens, "Editorial: The Importance of Raising the Profile of Alcohol Hangover Research," *Current Drug Abuse Reviews* 3, no. 2 (2010): 64.

Behind it is a pint of beer: Ibid., 66.

twelve to fourteen hours later: Renske Penning et al., "The Pathology of Alcohol Hangover," *Current Drug Abuse Reviews* 3, no. 2 (2010): 69.

190 *hangovers are usually the opposite:* Wiese, Shlipak, and Browner, "The Alcohol Hangover," 900.

hangovers occur after just one session: Penning, "Pathology of Alcohol Hangover," 68.

people who are less sensitive: Jonathan Howland et al., "Proceedings of the 2010 Symposium on Hangover and Other Residual Alcohol Effects: Predictors and Consequences," *Open Addiction Journal* 3 (2010): 131.

larger region of the cortex lit up: Ibid., 132.

192 *But the bourbon drinkers reported* worse *hangovers:* Damaris J. Rohsenow and Jonathan Howland, "The Role of Beverage Congeners in Hangover and Other Residual Effects of Alcohol Intoxication: A Review," *Current Drug Abuse Reviews* 3, no. 2 (2010): 77.

turns methanol into formaldehyde: Gemma Prat, Ana Adan, and Miquel Sa, "Alcohol Hangover: A Critical Review of Explanatory Factors," *Human Psychopharmacology* 24 (April 2009): 259–67.

193 *Formic acid or formate:* James A. Sharpe et al., "Methanol Optic Neuropathy: A Histopathological Study," *Neurology* 32, no. 10 (October 1, 1982): 1099.

The enzyme goes to work: J. D. Pritchard, *Methanol Toxicological Overview,* Health Protection Agency, 2007.

The patient just pees the methanol: Rohsenow and Howland, "Role of Beverage Congeners," 77.

194 *more than one in ten social drinkers:* Joris C. Verster and Renske Penning, "Treatment and Prevention of Alcohol Hangover," *Current Drug Abuse Reviews* 3, no. 2 (2010): 108.

One research team in Korea: Dai-Jin Kim et al., "Effects of Alcohol Hangover on Cytokine Production in Healthy Subjects," *Alcohol* 31, no. 3 (November 2003): 167–70.

higher-than-normal cytokine levels: Joris C. Verster, "The Alcohol Hangover — A Puzzling Phenomenon," *Alcohol and Alcoholism* 43, no. 2 (2008): 124.

195 *It keeps you from getting drunk:* Yi Shen et al., "Dihydromyricetin as a Novel Anti-alcohol Intoxication Medication," *Journal of Neuroscience* 32, no. 1 (January 4, 2012): 390.

196 *the ethanol target everyone is hunting for:* Hong Nie et al., "Extrasynaptic Delta-containing $GABA_A$ Receptors in the Nucleus Accumbens Dorsomedial Shell Contribute to Alcohol Intake," *Proceedings of the National Academy of Sciences of the United States of America* 108, no. 11 (March 15, 2011): 4459.

199 *subjected them to scientific tests:* Verster and Penning, "Treatment and Prevention," 108.

a group of researchers in Finland: S. Kaivola et al., "Hangover Headache and Prostaglandins: Prophylactic Treatment with Tolfenamic Acid," *Cephalalgia* 3, no. 1 (March 1983): 31–36.

extract of the skin of the prickly pear: Jeffrey Wiese and S. McPherson, "Effect of *Opuntia ficus indica* on Symptoms of the Alcohol Hangover," *Archives of Internal Medicine* 164 (2004): 1334.

CONCLUSION

205 *"cultural meanings and roles of alcohol":* Social Issues Research Centre, *Social and Cultural Aspects of Drinking* (Oxford, UK: Social Issues Research Centre, 1998), 26.

206 *much more harmful than marijuana:* David J. Nutt, Leslie A. King, and Lawrence D. Phillips, "Drug Harms in the UK: A Multicriteria Decision Analysis," *Lancet* 376, no. 9752 (November 6, 2010): 1561.

drunk driving caused over 13,000 deaths: Matthew Chambers, Mindy Liu, and Chip Moore, "Drunk Driving by the Numbers," US Department of Transportation, http://www.rita.dot.gov/bts/sites/rita.dot.gov.bts/files/publications/by_the_numbers/drunk_driving/index.html.

emergency room intake records: B. Taylor et al., "The More You Drink, the Harder You Fall: A Systematic Review and Meta-analysis of How Acute Alcohol Consumption and Injury or Collision Risk Increase Together," *Drug and Alcohol Dependence* 110 (July 1, 2010): 115.

207 *the vets they served as "family":* Keith A. Anderson, Jeffrey J. Maile, and Lynette G. Fisher, "The Healing Tonic: A Pilot Study of the Perceived Ability and Potential of Bartenders," *Journal of Military and Veterans' Health* 18, no. 4 (2010): 17.

those pesky giant pandas: Lee Hannah et al., "Climate Change, Wine, and Conservation," *Proceedings of the National Academy of Sciences* 110, no. 17 (2013): 6910.

208 *or cure a hangover:* David J. Nutt, "Alcohol Alternatives — A Goal for Psychopharmacology?" *Journal of Psychopharmacology* 20, no. 3 (2006): 318.

Nutt wrote in a commentary: David J. Nutt, "Alcohol Without the Hangover? It's Closer than You Think," *Shortcuts* (blog), *Guardian* (November 11, 2013), http://www.theguardian.com/commentisfree/2013/nov/11/alcohol-benefits-no-dangers-closer-thinkhttp://www.theguardian.com/commentisfree/2013/nov/11/alcohol-benefits-no-dangers-closer-think.

AFTERWORD

214 *It's not a crowded field:* Michael Kinstlick. "The US Craft Distilling Market: 2013 Update," Coppersea Distilling (2013), 2.

215 *A significant portion:* Eric Felten, "Your 'Craft' Whiskey Is Probably from a Factory Distillery in Indiana," *Daily Beast,* July 28, 2014. Accessed December 7, 2014. http://www.thedailybeast.com/articles/2014/07/28/your-craft-whiskey-is-probably-from-a-factory-distillery-in-indiana.html.

It'd be as if a giant: Anna Toon, "Celis Brewery, Back Home Where It Belongs," *Austin Chronicle,* Nov. 1, 2013. Accessed December 27, 2014. http://www.austinchronicle.com/food/2013-11-01/celis-brewery-back-home-where-it-belongs. Also see Esteban On, "12 Craft Beers That Aren't Really Craft

Beers," *Refined Guy.* Accessed December 27, 2014. http://www.refinedguy.com/2013/03/07/craft-beers-that-arent-really-craft-beers.

216 *In fact—as the lawsuit:* Josh Hafner, "How Templeton Rye Is Produced," *Des Moines Register,* August 28, 2014. Accessed December 7, 2014. http://www.desmoinesregister.com/story/news/2014/08/28/templeton-rye-produced/14770031.

In an interview with a big-shot: "The Templeton Tale Gets Curioser and Curioser," Chuck Cowdery blog, October 3, 2014. http://chuckcowdery.blogspot.com/2014/10/templeton-tale-gets-curiouser-and.html.

217 *If you're willing to get:* http://www.professoren.tum.de/en/hofmann-thomas.

In a series of papers: Nadine Wollmann, Jan Carlos Hufnagel, and Thomas Hofmann, "Decoding the Taste of Red Wine Using a Sensomics Approach," in *Flavour Science: Proceedings from XIII Weurman Flavour Research Symposium,* ed. Vicente Ferreira and Ricardo Lopez (San Diego: Elsevier, 2014), 525.

Over the past half-dozen: Lauren K. Wolf, "A Taste of Wine Science," *Chemical and Engineering News* (September 22, 2014), 28.

Anyway, science already offers: Christopher Null, "How to Make Mass-Produced Wine Taste Great," *Wired,* May 2013. Accessed December 7, 2014. http://www.wired.com/2014/04/how-to-make-wine-taste-good.

BIBLIOGRAPHY

Adams, David J. "Fungal Cell Wall Chitinases and Glucanases." *Microbiology* 150, part 7 (2004): 2029–35.

Adams, Douglas. *The Hitchhiker's Guide to the Galaxy.* New York: Harmony Books, 1979.

Akiyama, Hiroichi. *Saké: The Essence of 2000 Years of Japanese Wisdom Gained from Brewing Alcoholic Beverages from Rice.* Translated by Inoue Takashi. Tokyo: Brewing Society of Japan, 2010.

Allchin, F. R. "India: The Ancient Home of Distillation?" *Man* 14, no. 1 (1979): 55–63.

Anderson, Keith A., Jeffrey J. Maile, and Lynette G. Fisher. "The Healing Tonic: A Pilot Study of the Perceived Ability and Potential of Bartenders." *Journal of Military and Veterans' Health* 18, no. 4 (2010): 17–24.

Antonow, David R., and Craig J. McClain. "Nutrition and Alcoholism." In *Alcohol and the Brain: Chronic Effects,* edited by Ralph Tarter and David Van Thiel, 81–120. New York: Plenum Medical Book Company, 1985.

Arnold, Wilfred Niels. "Absinthe." *Scientific American,* June 1989. http://www.scientificamerican.com/article.cfm?id=absinthe-history.

Arroyo, Rafael. *Studies on Rum.* Research Bulletin no. 5. Rio Piedras: University of Puerto Rico Agricultural Experimental Station, December 1945.

Arroyo-García, R., L. Ruiz-García, L. Bolling, R. Ocete, M. A. López, C. Arnold, A. Ergul, et al. "Multiple Origins of Cultivated Grapevine (*Vitis Vinifera* L. Ssp. *Sativa*) Based on Chloroplast DNA Polymorphisms." *Molecular Ecology* 15 (October 2006): 3707–14.

Ashcraft, Brian. "The Mystery of the Green Menace." *Wired,* November 2005.

Atkins, Peter W. *Molecules.* New York: Scientific American Library, 1987.

Atkinson, R. W. *The Chemistry of Sake Brewing.* Tokyo: Tokyo University, 1881.

Attwood, Angela S., Nicholas E. Scott-Samuel, George Stothart, and Marcus R. Munafò. "Glass Shape Influences Consumption Rate for Alcoholic Beverages." *PLoS ONE* 7, no. 8 (January 2012): e43007.

Bachmanov, Alexander A., Stephen W. Kiefer, Juan Carlos Molina, Michael G. Tordoff, Valerie B. Duffy, Linda M. Bartoshuk, and Julie A. Mennella. "Chemosensory Factors Influencing Alcohol Perception, Preferences, and Consumption." *Alcoholism: Clinical and Experimental Research* 27, no. 2 (February 2003): 220–31.

Bamforth, Charles W. *Foam.* St. Paul, MN: American Society of Brewing Chemists, 2012.

———. *Scientific Principles of Malting and Brewing.* St. Paul, MN: American Society of Brewing Chemists, 2006.

Barnett, Brendon. "Fermentation." Pasteur Brewing. Modified December 29, 2011. http://www.pasteurbrewing.com/louis-pasteur-fermenation.html.

Barnett, James A., and Linda Barnett. *Yeast Research: A Historical Overview.* London: ASM Press, 2011.

Begleiter, Henri, and Arthur Platz. "The Effects of Alcohol on the Central Nervous System in Humans." In *The Biology of Alcoholism,* edited by B. Kissin and Henri Begleiter, 293–343. New York: Plenum Publishing Corporation, 1972.

Benedict, Francis G., and Raymond Dodge. *Psychological Effects of Alcohol; an Experimental Investigation of the Effects of Moderate Doses of Ethyl Alcohol on a Related Group of Neuro-muscular Processes in Man.* Washington, DC: Carnegie Institute of Washington, 1915.

Bennett, Joan W. "Adrenalin and Cherry Trees." *Modern Drug Discovery* 4, no. 12 (2001): 47–48. http://pubs.acs.org/subscribe/archive/mdd/v04/i12/html/12timeline.html.

———. Untitled presentation at the 2012 American Society for Microbiology Meeting, San Francisco, June 17, 2012.

Bokulich, Nicholas A., John H. Thorngate, Paul M. Richardson, and David A. Mills. "Microbial Biogeography of Wine Grapes Is Conditioned by Cultivar, Vintage, and Climate." *Proceedings of the National Academies of Science* (published online ahead of print November 25, 2013): 1–10. http://www.pnas.org/content/early/2013/11/20/1317377110.full.pdf+html.

Boothby, William T. *Cocktail Boothby's American Bar-Tender.* San Francisco: H. S. Crocker, 1891. Reprinted with a foreword by David Burkhart. San Francisco: Anchor Distilling, 2008.

Bouby, Laurent, Isabel Figueiral, Anne Bouchette, Nuria Rovira, Sarah Ivorra, Thierry Lacombe, Thierry Pastor, Sandrine Picq, Philippe Marinval, and Jean-Frédéric Terral. "Bioarchaeological Insights into the Process of Domestication of Grapevine (*Vitis vinifera* L.) During Roman Times in Southern France." *PLoS ONE* 8 (May 15, 2013): e63195.

Buemann, Benjamin, and Arne Astrup. "How Does the Body Deal with Energy from Alcohol?" *Nutrition* 17 (2001): 638–41.

Buffalo Trace Distillery. *Warehouse X* (blog). http://www.experimentalwarehouse.com.

Chambers, Matthew, Mindy Liu, and Chip Moore. "Drunk Driving by the Numbers." United States Department of Transportation. http://www.rita.dot.gov/bts/sites/rita.dot.gov.bts/files/publications/by_the_numbers/drunk_driving/index.html.

Chandrashekar, Jayaram, David Yarmolinsky, Lars von Buchholtz, Yuki Oka, William Sly, Nicholas J. P. Ryba, and Charles S. Zuker. "The Taste of Carbonation." *Science* 326, no. 5951 (October 16, 2009): 443–45.

Chemical Heritage Foundation. "Justus von Liebig and Friedrich Wöhler." http://www.chemheritage.org/discover/online-resources/chemistry-in-history/themes/molecular-synthesis-structure-and-bonding/liebig-and-wohler.aspx.

Clarke, Hewson, and John Dougall. *The Cabinet of Arts, or General Instructor in Arts, Science, Trade, Practical Machinery, the Means of Preserving Human Life, and Political Economy, Embracing a Variety of Important Subjects.* London: T. Finnersley, 1817.

Conison, Alexander. "The Organization of Rome's Wine Trade." PhD diss., University of Michigan, 2012. http://deepblue.lib.umich.edu/bitstream/handle/2027.42/91455/conison_1.pdf?sequence=1.

Correa, Mercè, John D. Salamone, Kristen N. Segovia, Marta Pardo, Rosanna Longoni, Liliana Spina, Alessandra T. Peana, Stefania Vinci, and Elio Acquas. "Piecing Together the Puzzle of Acetaldehyde as a Neuroactive Agent." *Neuroscience and Biobehavioral Reviews* 36, no. 1 (January 2012): 404–30.

Court of Master Sommeliers. "Courses & Schedules." http://www.mastersommeliers.org/Pages.aspx/Master-Sommelier-Diploma-Exam.

Cowdery, Charles K. *Bourbon, Straight: The Uncut and Unfiltered Story of American Whiskey.* Chicago: Made and Bottled in Kentucky, 2004.

Debré, Patrice. *Louis Pasteur.* Translated by Elborg Forster. Baltimore: Johns Hopkins University Press, 1994.

De Keersmaecker, Jacques. "The Mystery of Lambic Beer." *Scientific American,* August 1996.

Delwiche, J. F., and Marcia Levin Pelchat. "Influence of Glass Shape on Wine Aroma." *Journal of Sensory Studies* 17, no. 2002 (2002): 19–28.

Dietrich, Oliver, Manfred Heun, Jens Notroff, Klaus Schmidt, and Martin Zarnkow. "The Role of Cult and Feasting in the Emergence of Neolithic Communities. New Evidence from Gobekli Tepe, South-eastern Turkey." *Antiquity* 86 (2012): 674–95.

Dogfish Head Brewing. "Midas Touch." http://www.dogfish.com/brews-spirits/the-brews/year-round-brews/midas-touch.htm.

Dressler, David, and Huntington Porter. *Discovering Enzymes.* New York: Scientific American Library, 1991.

Dudley, Robert. "Evolutionary Origins of Human Alcoholism in Primate Frugivory." *Quarterly Review of Biology* 75, no. 1 (2000): 3–15.

Dunn, Barbara, and Gavin Sherlock. "Reconstruction of the Genome Origins and Evolution of the Hybrid Lager Yeast *Saccharomyces pastorianus.*" *Genome Research* 18, no. 650 (2008): 1610–23.

E. C. "On the Antiquity of Brewing and Distillation in Ireland." *Ulster Journal of Archaeology* 7 (1859): 33–40.

Eng, Mimy Y., Susan E. Luczak, and Tamara L. Wall. "ALDH2, ADH1B, and ADH1C Genotypes in Asians: A Literature Review." *Alcohol Research & Health* 30, no. 1 (2007): 22–27.

Epstein, Murray. "Alcohol's Impact on Kidney Function." *Alcohol Health & Research World* 21, no. 1 (1997): 84–93.

Fay, Justin C., and Joseph A. Benavides. "Evidence for Domesticated and Wild Populations of *Saccharomyces cerevisiae*." *PLoS Genetics* 1 (2005): 66–71.

Forbes, R. J. *Short History of the Art of Distillation*. Leiden, the Netherlands: E. J. Brill, 1948.

Geison, Gerald L. *The Private Science of Louis Pasteur*. Princeton: Princeton University Press, 1995.

Gergaud, Olivier, and Victor Ginsburgh. "Natural Endowments, Production Technologies and the Quality of Wines in Bordeaux. Does Terroir Matter?" *Journal of Wine Economics* 5 (2010): 3–21.

Gevins, Alan, Cynthia S. Chan, and Lita Sam-Vargas. "Towards Measuring Brain Function on Groups of People in the Real World." *PLoS ONE* 7, no. 9 (September 5, 2012): e44676.

Gibbons, John G., Leonidas Salichos, Jason C. Slot, David C. Rinker, Kriston L. McGary, Jonas G. King, Maren A. Klich, David L. Tabb, W. Hayes McDonald, and Antonis Rokas. "The Evolutionary Imprint of Domestication on Genome Variation and Function of the Filamentous Fungus *Aspergillus oryzae*." *Current Biology* 22 (2012): 1–7.

Goffeau, A., B. G. Barrell, H. Bussey, R. W. Davis, B. Dujon, H. Feldmann, F. Galibert, et al. "Life with 6000 Genes." *Science* 274 (1996): 546–67.

Goldman, Jason G. "Dogs, but Not Wolves, Use Humans as Tools." *The Thoughtful Animal* (blog), *Scientific American*, April 30, 2012. http://blogs.scientificamerican.com/thoughtful-animal/2012/04/30/dogs-but-not-wolves-use-humans-as-tools/.

Goode, Jaime. *The Science of Wine*. Berkeley: University of California Press, 2005.

Granados, J. Quesada, J. J. Merelos Guervós, M. J. Olveras López, J. Gonzales Peñalver, M. Olalla Herrera, R. Blanca Herrera, and M. C. Lopez Martinez. "Application of Artificial Aging Techniques to Samples of Rum and Comparison with Traditionally Aged Rums by Analysis with Artificial Neural Nets." *Journal of Agricultural and Food Chemistry* 50, no. 6 (March 13, 2002): 1470–77.

Gray, W. Blake. "Bacardi, and Its Yeast, Await a Return to Cuba." *Los Angeles Times*, October 6, 2011. http://latimes.com/features/food/la-fo-bacardi-20111006,0,1042.story.

Gross, Leonard. *How Much Is Too Much? The Effects of Social Drinking*. New York: Magilla, 1983.

Haag, H. B., J. K. Finnegan, P. S. Larson, and R. B. Smith. "Studies on the Acute Toxicity and Irritating Properties of the Congeners in Whisky." *Toxicology and Applied Pharmacology* 627, no. 6 (1959): 618–27.

Hackbarth, James J. "Multivariate Analyses of Beer Foam Stand." *Journal of the Institute of Brewing* 112, no. 1 (2006): 17–24.

Hannah, Lee, Patrick R. Roehrdanz, Makihiko Ikegami, Anderson V. Shepard, M. Rebecca Shaw, Gary Tabor, Lu Zhi, Pablo A. Marquet, and Robert J. Hijmans. "Climate Change, Wine, and Conservation." *Proceedings of the National Academy of Sciences* (2013): 2–7.

Harrison, Barry, Olivier Fagnen, Frances Jack, and James Brosnan. "The Impact of Copper in Different Parts of Malt Whisky Pot Stills on New Make Spirit Composition and Aroma." *Journal of the Institute of Brewing* 117, no. 1 (2011): 106–12.

Harrison, Mark A. "Beer/Brewing." In *Encyclopedia of Microbiology,* 3rd ed., edited by Moselio Schaechter, 23–33. San Diego: Academic Press, 2009.

Hevesi, Dennis. "G. A. Marlatt, Advocate of Shift in Treating Addicts, Dies at 69." *New York Times,* March 21, 2011. http://www.nytimes.com/2011/03/22/us/22marlatt.html?_r=0.

Hilgard, E. R. *Walter Richard Miles, 1885–1978.* Washington DC: National Academy of Sciences, 1985. http://www.nasonline.org/publications/biographical-memoirs/memoir-pdfs/miles-walter.pdf.

Himwich, Harold E. "The Physiology of Alcohol." *Journal of the American Medical Association* 1446, no. 7 (1957): 545–49.

Hornsey, Ian. *Alcohol and Its Role in the Evolution of Human Society.* Cambridge: Royal Society of Chemistry, 2012.

———. *The Chemistry and Biology of Winemaking.* Cambridge: Royal Society of Chemistry, 2007.

———. *A History of Beer and Brewing.* Cambridge: Royal Society of Chemistry, 2004.

Hough, James S. *The Biotechnology of Malting and Brewing.* Cambridge: Cambridge University Press, 1985.

Howland, Jonathan, Damaris J. Rohsenow, John E. McGeary, Chris Streeter, and Joris C. Verster. "Proceedings of the 2010 Symposium on Hangover and Other Residual Alcohol Effects: Predictors and Consequences." *Open Addiction Journal* 3 (2010): 131–32.

Hu, Naiping, Dan Wu, Kelly Cross, Sergey Burikov, Tatiana Dolenko, Svetlana Patsaeva, and Dale W. Schaefer. "Structurability: A Collective Measure of the Structural Differences in Vodkas." *Journal of Agricultural and Food Chemistry* 58 (2010): 7394–401.

Huang, H. T. *Science and Civilisation in China.* Vol. 6, pt. 5, *Biology and Biological Technology: Fermentations and Food.* Cambridge: Cambridge University Press, 2000.

Hummel, T., J. F. Delwiche, C. Schmidt, and K.-B. Hüttenbrink. "Effects of the Form of Glasses on the Perception of Wine Flavors: A Study in Untrained Subjects." *Appetite* 41, no. 2 (October 2003): 197–202.

Independent Stave Company. *International Barrel Symposium: Research Results and Highlights from the 5th, 6th, and 7th Symposiums.* Lebanon, MO: Independent Stave Company, 2008.

Ingham, Richard. "Champagne Physicist Reveals the Secrets of Bubbly." *Phys.org.* Last modified September 18, 2012. http://phys.org/news/2012-09-champagne-physicist-reveals-secrets.html.

Isaacson, Walter. *Benjamin Franklin: An American Life.* New York: Simon & Schuster, 2003.

Jayarajah, Christine N., Alison M. Skelley, Angela D. Fortner, and Richard A. Mathies. "Analysis of Neuroactive Amines in Fermented Beverages Using a Portable Microchip Capillary Electrophoresis System." *Analytical Chemistry* 79, no. 21 (November 2007): 8162–69.

Johnson, Keith, David B. Pisoni, and Robert H. Bernacki. "Do Voice Recordings Reveal Whether a Person Is Intoxicated? A Case Study." *Phonetica* 47 (1990): 215–37.

Kaivola, S., J. Parantainen, T. Osterman, and H. Timonen. "Hangover Headache and Prostaglandins: Prophylactic Treatment with Tolfenamic Acid." *Cephalalgia* 3, no. 1 (March 1983): 31–36.

Katz, Sandor. *The Art of Fermentation.* White River Junction, VT: Chelsea Green Publishing, 2012.

Kawakami, K. K. *Jokichi Takamine: A Record of His American Achievements.* New York: William Edwin Rudge, 1928.

Kaye, Joseph N. "Symbolic Olfactory Display." PhD diss., Massachusetts Institute of Technology, 2001.

Kim, Dai-Jin, Won Kim, Su-Jung Yoon, Bo-Moon Choi, Jung-Soo Kim, Hyo Jin Go, Yong-Ku Kim, and Jaeseung Jeong. "Effects of Alcohol Hangover on Cytokine Production in Healthy Subjects." *Alcohol* 31, no. 3 (November 2003): 167–70.

King, Ellena S., Randall L. Dunn, and Hildegarde Heymann. "The Influence of Alcohol on the Sensory Perception of Red Wines." *Food Quality and Preference* 28, no. 1 (April 2013): 235–43.

Kintslick, Michael. *The U.S. Craft Distilling Market: 2011 and Beyond.* New York: Coppersea Distillery, 2011.

Kitamoto, Katsuhiko. "Molecular Biology of the Koji Molds." *Advances in Applied Microbiology* 51 (January 2002): 129–53.

Lachenmeier, D. W., David Nathan-Maister, Theodore A. Breaux, Jean-Pierre Luauté, and Emmert Joachim. "Absinthe, Absinthism and Thujone—New Insight into the Spirit's Impact on Public Health." *Open Addiction Journal* 3 (2010): 32–38.

Lagi, Marco, and R. S. Chase. "Distillation: Integration of a Historical Perspective." *Australian Journal of Education in Chemistry* 70 (2009): 5–10.

Leffingwell, John. "Update No. 5: Olfaction." *Leffingwell Reports* 2 (May 2002).

Lehrer, Adrienne. *Wine & Conversation.* 2nd ed. New York: Oxford University Press, 2009.

Libkind, D., C. T. Hittinger, E. Valerio, C. Goncalves, J. Dover, M. Johnston, P. Goncalves, and J. P. Sampaio. "Microbe Domestication and the Identification of the Wild Genetic Stock of Lager-brewing Yeast." *Proceedings of the National Academy of Sciences* 108, no. 34 (August 22, 2011).

Liger-Belair, Gérard. *Uncorked: The Science of Champagne.* Princeton: Princeton University Press, 2004.

Liger-Belair, Gérard, Clara Cilindre, Régis D. Gougeon, Marianna Lucio, Istvan Gebefügi, Philippe Jeandet, and Philippe Schmitt-Kopplin. "Unraveling Different Chemical Fingerprints Between a Champagne Wine and Its Aerosols." *Proceedings of the National Academies of Science* 106, no. 39 (2009): 16545–49.

Liger-Belair, Gérard, Guillaume Polidori, and Philippe Jeandet. "Recent Advances in the Science of Champagne Bubbles." *Chemical Society Reviews* 37, no. 11 (November 2008): 2490–511.

Liu, KeShun. "Chemical Composition of Distillers Grains, a Review." *Journal of Agricultural and Food Chemistry* 59 (March 9, 2011): 1508–26.

Livermore, Andrew, and David G. Laing. "The Influence of Chemical Complexity on the Perception of Multicomponent Odor Mixtures." *Perception & Psychophysics* 60, no. 4 (May 1998): 650–61.

Lund, Steven T., and Joerg Bohlmann. "The Molecular Basis for Wine Grape Quality—A Volatile Subject." *Science* 311 (February 10, 2006): 804–5.

MacAndrew, Craig, and Robert Edgerton. *Drunken Comportment: A Social Explanation.* Hawthorn, NY: Aldine, 1969. Reprinted with a foreword by Dwight B. Heath. Clinton Corners, NY: Elliot Werner, 2003.

Macatelli, Melina, John R. Piggott, and Alistair Paterson. "Structure of Ethanol-Water Systems and Its Consequences for Flavour." In *New Horizons: Energy, Environment, and Enlightenment: Proceedings of the Worldwide Distilled Spirits Conference,* edited by G. M. Walker and P. S. Hughes, 235–42. Nottingham: Nottingham University Press, 2010.

Machida, Masayuki, Kiyoshi Asai, Motoaki Sano, Toshihiro Tanaka, Toshitaka Kumagai, Goro Terai, Ken-Ichi Kusumoto, et al. "Genome Sequencing and Analysis of *Aspergillus oryzae.*" *Nature* 438, no. 7071 (December 22, 2005): 1157–61.

Machida, Masayuki, Osamu Yamada, and Katsuya Gomi. "Genomics of *Aspergillus oryzae*: Learning from the History of Koji Mold and Exploration of Its Future." *DNA Research* 15 (August 2008): 173–83.

Maisto, Stephen A., Gerard J. Connors, and Paul R. Sachs. "Expectation as a Mediator in Alcohol Intoxication: A Reference Level Model." *Cognitive Therapy and Research* 5, no. 1 (1981): 1–18.

Malnic, Bettina, Daniela C. Gonzalez-Kristeller, and Luciana M. Gutiyama. "Odorant Receptors." In *The Neurobiology of Olfaction,* edited by Anna Menini, 181–202. Boca Raton, FL: CRC Press, 2010.

Marlatt, G. Alan, Barbara Demming, and John B. Reid. "Loss of Control Drinking in Alcoholics: An Experimental Analogue." *Journal of Abnormal Psychology* 81, no. 3 (1973): 233–41.

———. "This Week's Citation Classic." *ISI Current Contents: Social and Behavioral Sciences* 18 (1985): 18.

Marshall, K., David G. Laing, A. L. Jinks, and I. Hutchinson. "The Capacity of Humans to Identify Components in Complex Odor-Taste Mixtures." *Chemical Senses* 31, no. 6 (July 2006): 539–45.

McGovern, Patrick E. *Ancient Wine: The Search for the Origins of Viniculture.* Princeton: Princeton University Press, 2007.

McGovern, Patrick E., Juzhong Zhang, Jigen Tang, Zhiqing Zhang, Gretchen R. Hall, Robert A. Moreau, Alberto Nunez, et al. "Fermented Beverages of Pre- and Proto-historic China." *Proceedings of the National Academy of Sciences* 101, no. 51 (December 2004): 17593–98.

McKenzie, Judith. *The Architecture of Alexandria and Egypt 300 BC–AD 700.* New Haven: Yale University Press, 2007.

Meier, Sebastian, Magnus Karlsson, Pernille R. Jensen, Mathilde H. Lerche, and Jens Ø. Duus. "Metabolic Pathway Visualization in Living Yeast by DNP-NMR." *Molecular bioSystems* 7 (October 2011): 2834–36.

Menz, Garry, Christian Andrighetto, Angiolella Lombardi, Viviana Corich, Peter Aldred, and Frank Vriesekoop. "Isolation, Identification, and Characterisation of Beer-spoilage Lactic Acid Bacteria from Microbrewed Beer from Victoria, Australia." *Journal of the Institute of Brewing* 116 (2010): 14–22.

Michel, Rudolph H., Patrick E. McGovern, and Virginia R. Badler. "The First Wine & Beer: Chemical Detection of Ancient Fermented Beverages." *Analytical Chemistry* 65, no. 8 (April 1993): 408A–13A.

Miles, W. R. *Alcohol and Human Efficiency.* Washington, DC: Carnegie Institute of Washington, 1924.

Mitchell, Jennifer M., James P. O'Neil, Mustafa Janabi, Shawn M. Marks, William J. Jagust, and Howard L. Fields. "Alcohol Consumption Induces Endogenous Opioid Release in the Human Orbitofrontal Cortex and Nucleus Accumbens." *Science Translational Medicine* 4, no. 116 (January 11, 2012): 116ra6.

Moran, Bruce. *Distilling Knowledge: Alchemy, Chemistry, and the Scientific Revolution.* Cambridge: Harvard University Press, 2005.

Morrot, Gil, Frédéric Brochet, and Denis Dubourdieu. "The Color of Odors." *Brain and Language* 79, no. 2 (2001): 309–20.

Mosedale, J. R., and Jean-Louis Puech. "Barrels: Wines, Spirits, and Other Beverages." In *Encyclopedia of Food Sciences and Nutrition,* edited by Benjamin Caballero, Luiz C. Trugo, and Paul M. Finglass, 393–403. San Diego: Academic Press, 2003.

———. "Wood Maturation of Distilled Beverages." *Trends in Food Science & Technology* 9, no. 3 (March 1998): 95–101.

Murray, Jim. "Tomorrow's Malt." *Whisky* 1 (1999): 56.

Murtagh, John E. "Feedstocks, Fermentation and Distillation for Production of Heavy and Light Rums." In *The Alcohol Textbook: A Reference for the Beverage, Fuel and Industrial Alcohol Industries,* edited by K. A. Jacques, T. P. Lyons, and D. R. Kelsall, 243–55. Nottingham: Nottingham University Press, 1999.

National Library of Medicine. "Aspergillosis." Last modified May 19, 2013. http://www.ncbi.nlm.nih.gov/pubmedhealth/PMH0002302/.

Negrul, A. M. "Method and Apparatus for Harvesting Grapes." US Patent 3,564,827. Washington, DC: US Patent and Trademark Office, 1971.

Nicol, D. "Batch Distillation." In *The Science and Technology of Whiskies,* edited by J. R. Piggott, R. Sharp, and R. E. B. Duncan, 118–49. London: Longman Group, 1989.

Nie, Hong, Mridula Rewal, T. Michael Gill, Dorit Ron, and Patricia H. Janak. "Extrasynaptic Delta-containing $GABA_A$ Receptors in the Nucleus Accumbens Dorsomedial Shell Contribute to Alcohol Intake." *Proceedings of the National Academy of Sciences of the United States of America* 108, no. 11 (March 15, 2011): 4459–64.

Noll, Roger G. "The Wines of West Africa: History, Technology and Tasting Notes." *Journal of Wine Economics* 3 (2008): 85–94.

Nutt, David J. "Alcohol Alternatives — A Goal for Psychopharmacology?" *Journal of Psychopharmacology* 20, no. 3 (May 2006): 318–20.

———. "Alcohol Without the Hangover? It's Closer than You Think." *Shortcuts* (blog), *Guardian,* November 11, 2013. http://www.theguardian.com/commentisfree/2013/nov/11/alcohol-benefits-no-dangers-closer-think.

Nutt, David J., Leslie A. King, and Lawrence D. Phillips. "Drug Harms in the UK: A Multicriteria Decision Analysis." *Lancet* 376, no. 9752 (November 6, 2010): 1558–65.

Oakes, Elizabeth H. *Encyclopedia of World Scientists.* New York: Infobase Learning, 2007.

Our Knowledge Box: Or, Old Secrets and New Discoveries. New York: Geo. Blackie and Co., 1875.

Palmer, Geoff H. "Beverages: Distilled." In *Encyclopedia of Grain Science,* edited by Colin Wrigley, Harold Corke, and Charles E. Walker, 96–108. San Diego: Academic Press, 2004.

Panconesi, Alessandro. "Alcohol and Migraine: Trigger Factor, Consumption, Mechanisms: A Review." *Journal of Headache and Pain* 9, no. 1 (2008): 19–27.

Panek, Richard J., and Armond R. Boucher. "Continuous Distillation." In *The Science and Technology of Whiskies,* edited by J. R. Piggott, R. Sharp, and R. E. B. Duncan, 150–81. London: Longman Group, 1989.

Patai, Raphael. *The Jewish Alchemists.* Princeton: Princeton University Press, 1994.

"Peat and Its Products." *Illustrated Magazine of Art* 1 (1953): 374–75.

Pelchat, Marcia Levin, and Fritz Blank. "A Scientific Approach to Flavours and Olfactory Memory." In *Food and the Memory: Proceedings of the Oxford Sym-*

posium on Food and Cookery, edited by Harlan Walker, 185–91. Devon: Prospect Books, 2001.

Penning, Renske, Merel van Nuland, Lies A. L. Fliervoet, Berend Olivier, and Joris C. Verster. "The Pathology of Alcohol Hangover." *Current Drug Abuse Reviews* 3, no. 2 (2010): 68–75.

Phaff, Herman Jan, Martin W. Miller, and Emil M. Mrak. *The Life of Yeasts: Second Revised and Enlarged Edition.* Cambridge: Harvard University Press, 1966.

Philp, J. M. "Cask Quality and Warehouse Conditions." In *The Science and Technology of Whiskies,* edited by J. R. Piggott, R. Sharp, and R. E. B. Duncan, 273–74. London: Longman Group, 1989.

Piggot, Robert. "Beverage Alcohol Distillation." In *The Alcohol Textbook,* 5th ed., edited by W. M. Ingeldew, D. R. Kelsall, G. D. Austin, and C. Kluhspies, 431–43. Nottingham: Nottingham University Press, 2009.

———. "Rum: Fermentation and Distillation." In *The Alcohol Textbook,* 5th ed., edited by W. M. Ingeldew, D. R. Kelsall, G. D. Austin, and C. Kluhspies, 473–80. Nottingham: Nottingham University Press, 2009.

———. "Vodka, Gin and Liqueurs." In *The Alcohol Textbook,* 5th ed., edited by W. M. Ingeldew, D. R. Kelsall, G. D. Austin, and C. Kluhspies, 465–72. Nottingham: Nottingham University Press, 2009.

Pollard, Justin, and Howard Reid. *The Rise and Fall of Alexandria: Birthplace of the Modern Mind.* New York: Viking, 2006.

Prat, Gemma, Ana Adan, and Miquel Sa. "Alcohol Hangover: A Critical Review of Explanatory Factors." *Human Psychopharmacology* 24 (April 2009): 259–67.

Pretorius, Isak S., Christopher D. Curtin, and Paul J. Chambers. "The Winemaker's Bug: From Ancient Wisdom to Opening New Vistas with Frontier Yeast Science." *Bioengineered Bugs* 3, no. 3 (2012): 147–56.

Pritchard, J. D. *Methanol Toxicological Overview.* Chilton, Oxfordshire, UK: Health Protection Agency, 2007.

Quandt, R. E. "On Wine Bullshit: Some New Software?" *Journal of Wine Economics* 2, no. 2 (2007): 129–35.

Ratliff, Evan. "Taming the Wild." *National Geographic,* March 2011. http://ngm.nationalgeographic.com/2011/03/taming-wild-animals/ratliff-text.

Reardon, Sara. "Zebra Finches Sing Sloppily When Drunk." *New Scientist,* October 17, 2012. http://www.newscientist.com/article/dn22389-zebra-finches-sing-sloppily-when-drunk.html.

Reddy, Nischita K., Ashwani Singal, and Don W. Powell. "Alcohol-Related Diarrhea." In *Diarrhea: Diagnostic and Therapeutic Advances,* edited by Stefano Guandalini and Haleh Vaziri, 379–92. New York: Springer, 2011.

Richter, Chandra L., Barbara Dunn, Gavin Sherlock, and Tom Pugh. "Comparative Metabolic Footprinting of a Large Number of Commercial Wine Yeast Strains in Chardonnay Fermentations." *FEMS Yeast Research* 13, no. 4 (2013): 394–410.

Risen, Clay. "Whiskey Myth No. 2." *Mash Notes* (blog). Last modified July 27, 2012. http://clayrisen.com/?p=126.

Robinson, A. L., D. O. Adams, Paul K. Boss, H. Heymann, P. S. Solomon, and R. D. Trengove. "Influence of Geographic Origin on the Sensory Characteristics and Wine Composition of *Vitis vinifera* Cv. Cabernet Sauvignon Wines from Australia." *American Journal of Enology and Viticulture* 63, no. 4 (2012): 467–76.

Rodda, Luke N., Jochen Beyer, Dimitri Gerostamoulos, and Olaf H. Drummer. "Alcohol Congener Analysis and the Source of Alcohol: A Review." *Forensic Science, Medicine, and Pathology* 9, no. 2 (June 2013): 194–207.

Rodicio, Rosaura, and J. J. Heinisch. "Sugar Metabolism by Saccharomyces and Non-Saccharomyces Yeasts." In *Biology of Microorganisms on Grapes, in Must, and in Wine,* edited by H. König et al., 113–34. Berlin: Springer-Verlag, 2009.

Rohsenow, Damaris J., and Jonathan Howland. "The Role of Beverage Congeners in Hangover and Other Residual Effects of Alcohol Intoxication: A Review." *Current Drug Abuse Reviews* 3, no. 2 (2010): 76–79.

Roskrow, Dominic. "Is It the Age? Or the Mileage?" *Whisky Advocate* (Winter 2011): 77–80.

Rowley, Matthew. "Replacing That Worn Out Still — Every Ding and Dent?" *Rowley's Whiskey Forge* (blog). Last modified January 17, 2013. http://matthew-rowley.blogspot.com/2013/01/replacing-that-worn-out-still-every.html.

Scinska, Anna, Eliza Koros, Boguslaw Habrat, Andrzej Kukwa, Wojciech Kostowski, and Przemyslaw Bienkowski. "Bitter and Sweet Components of Ethanol Taste in Humans." *Drug and Alcohol Dependence* 60, no. 2 (August 1, 2000): 199–206.

Sharpe, James A., Michael Hostovsky, Juan M. Bilbao, and N. Barry Rewcastle. "Methanol Optic Neuropathy: A Histopathological Study." *Neurology* 32, no. 10 (October 1, 1982): 1093–1100.

Shen, Yi, A. Kerstin Lindemeyer, Claudia Gonzalez, Xuesi M. Shao, Igor Spigelman, Richard W. Olsen, and Jing Liang. "Dihydromyricetin as a Novel Anti-alcohol Intoxication Medication." *Journal of Neuroscience* 32, no. 1 (January 4, 2012): 390–401.

Shurtleff, William, and Akiko Aoyagi. *History of Koji — Grains and/or Soybeans Enrobed with a Mold Culture (300 BCE to 2012): Extensively Annotated Bibliography and Sourcebook.* Lafayette, CA: Soyinfo Center, 2012. http://www.soyinfocenter.com/pdf/154/Koji.pdf.

"The Singleton Distilleries: Glen Ord." *Whisky Advocate* (Spring 2013): 97.

"Singleton of Glen Ord," *Whisky News* (blog). http://malthead.blogspot.com/2006/12/singleton-of-glen-ord_10.html.

Sitnikova, N. L., Rudolf Sprik, Gerard Wegdam, and Erika Eiser. "Spontaneously Formed Trans-Anethol/Water/Alcohol Emulsions: Mechanism of Formation and Stability." *Langmuir* 21, no. 8 (2005): 7083–89.

Social Issues Research Centre. *Social and Cultural Aspects of Drinking.* Oxford, UK: Social Issues Research Centre, 1998.

Speers, R. Alex. "A Review of Yeast Flocculation." In *Yeast Flocculation, Vitality, and Viability: Proceedings of the 2nd International Brewers Symposium,* edited by R. Alex Speers, 1–16. St. Paul, MN: Master Brewers Association of the Americas, 2012.

Stajich, Jason E., Mary L. Berbee, Meredith Blackwell, David S. Hibbett, Timothy Y. James, Joseph W. Spatafora, and John W. Taylor. "The Fungi." *Current Biology* 19 (2009): R840–45.

Takamine, Jokichi. "Enzymes of Aspergillus Oryzae and the Application of Its Amyloclastic Enzyme to the Fermentation Industry." *Industrial & Engineering Chemistry* 6, no. 12 (1914): 824–28.

Taylor, B., H. M. Irving, F. Kanteres, Robin Room, G. Borges, C. J. Cherpitel, J. Bond, T. Greenfield, and J. Rehm. "The More You Drink, the Harder You Fall: A Systematic Review and Meta-analysis of How Acute Alcohol Consumption and Injury or Collision Risk Increase Together." *Drug and Alcohol Dependence* 110 (July 1, 2010): 108–16.

Taylor, Benjamin, and Jürgen Rehm. "Moderate Alcohol Consumption and Diseases of the Gastrointestinal System: A Review of Pathophysiological Processes." In *Alcohol and the Gastrointestinal Tract,* edited by Manfred Singer and David Brenner, 27–34. Basel: Karger Publishers, 2006.

Thompson, Derek. "The Economic Cost of Hangovers." *The Atlantic,* July 5, 2013. http://www.theatlantic.com/business/archive/2013/07/the-economic-cost-of-hangovers/277546/.

Thomson, J. Michael, Eric A. Gaucher, Michelle F. Burgan, Danny W. De Kee, Tang Li, John P. Aris, and Steven A. Benner. "Resurrecting Ancestral Alcohol Dehydrogenases from Yeast." *Nature Genetics* 37, no. 6 (June 2005): 630–35.

Tucker, Abigail. "The Beer Archaeologist." *Smithsonian,* July–August 2011. http://www.smithsonianmag.com/history-archaeology/The-Beer-Archaeologist.html?c=y&story=fullstory.

Vanderhaegen, B., H. Neven, H. Verachtert, and G. Derdelinckx. "The Chemistry of Beer Aging — A Critical Review." *Food Chemistry* 95, no. 3 (April 2006): 357–81.

Van Mulders, Sebastiaan, Luk Daenen, Pieter Verbelen, Sofie M. G. Saerens, Kevin J. Verstrepen, and Freddy R. Delvaux. "The Genetics Behind Yeast Flocculation: A Brewer's Perspective." In *Yeast Flocculation, Vitality, and Viability: Proceedings of the 2nd International Brewers Symposium,* edited by R. Alex Speers, 35–48. St. Paul, MN: Master Brewers Association of the Americas, 2012.

Verster, Joris C. "The Alcohol Hangover — A Puzzling Phenomenon." *Alcohol and Alcoholism* 43, no. 2 (2008): 124–26.

Verster, Joris C., and Renske Penning. "Treatment and Prevention of Alcohol Hangover." *Current Drug Abuse Reviews* 3, no. 2 (2010): 103–9.

Verster, Joris C., and Richard Stephens. "Editorial: The Importance of Raising the Profile of Alcohol Hangover Research." *Current Drug Abuse Reviews* 3, no. 2 (2010): 64–67.

Vrettos, Theodore. *Alexandria: City of the Western Mind.* New York: The Free Press, 2001.

"A Wake for Morten Christian Meilgaard." *Flower Parties through the Ages* (blog) http://goodfelloweb.com/flowerparty/fp_2009/Morten_Meilgaard_1928-2009.htm.

Wang, William Yang, Fadi Biadsy, Andrew Rosenberg, and Julia Hirschberg. "Automatic Detection of Speaker State: Lexical, Prosodic, and Phonetic Approaches to Level-of-interest and Intoxication Classification." *Computer Speech & Language* 27 (April 2012): 168–89.

Weiss, Tali, Kobi Snitz, Adi Yablonka, Rehan M. Khan, Danyel Gafsou, Elad Schneidman, and Noam Sobel. "Perceptual Convergence of Multi-Component Mixtures in Olfaction Implies an Olfactory White." *Proceedings of the National Academy of Sciences* 109, no. 49 (2012): 19959–64.

White, Chris, and Jamil Zainasheff. *Yeast: The Practical Guide to Beer Fermentation.* Boulder, CO: Brewers Publications, 2010.

White Labs. "About White Labs." Posted July 31, 2013. http://www.whitelabs.com/about_us.html.

White Labs. "Professional Yeast Bank." Posted July 31, 2013. http://www.whitelabs.com/beer/craft_strains.html.

Wiese, Jeffrey G., and S. McPherson. "Effect of *Opuntia ficus indica* on Symptoms of the Alcohol Hangover." *Archives of Internal Medicine* 164 (2004): 1334–40.

Wiese, Jeffrey G., Michael G. Shlipak, and Warren S. Browner. "The Alcohol Hangover." *Annals of Internal Medicine* 132, no. 11 (2000): 897–902.

Wilson, C. Anne. *Water of Life: A History of Wine-Distilling and Spirits 500 BC to AD 2000.* Devon, UK: Prospect Books, 2006.

Wilson, Donald A., and Robert L. Rennaker. "Cortical Activity Evoked by Odors." In *The Neurobiology of Olfaction,* edited by Anna Menini, 353–66. Boca Raton, FL: CRC Press, 2010.

Wilson, Donald A., and Richard J. Stevenson. *Learning to Smell: Olfactory Perception from Neurobiology to Behavior.* Baltimore: Johns Hopkins University Press, 2006.

Wood, Daniel. "Bar Lab Challenges the Alcohol Mystique." *Chicago Tribune,* February 24, 1991. http://articles.chicagotribune.com/1991-02-24/features/9101170848_1_addictive-behaviors-research-center-alcohol-free-alcoholism-and-alcohol-abuse.

Young, Emma. "Silent Song." *New Scientist,* October 27, 2000. http://www.newscientist.com/article/dn110-silent-song.html.

Zakhari, Samir. "Overview: How Is Alcohol Metabolized by the Body?" *Alcohol Research & Health* 29, no. 4 (January 2006): 245–54.

Zielinski, Sarah. "Hypatia, Ancient Alexandria's Great Female Scholar." *Smithson-*

ian.com. Last modified March 15, 2010. http://www.smithsonianmag.com/history-archaeology/Hypatia-Ancient-Alexandrias-Great-Female-Scholar.html.

Zucco, Gesualdo M., Aurelio Carassai, Maria Rosa Baroni, and Richard J. Stevenson. "Labeling, Identification, and Recognition of Wine-relevant Odorants in Expert Sommeliers, Intermediates, and Untrained Wine Drinkers." *Perception* 40, no. 5 (2011): 598–607.

INDEX